PROOFS THAT LIFE IS COSMIC

Acceptance of
a New Paradigm

PROOFS THAT LIFE IS COSMIC

Acceptance of a New Paradigm

Chandra Wickramasinghe

University of Buckingham, UK

Including the 1982 article *Proofs that Life is Cosmic* by Fred Hoyle & Chandra Wickramasinghe
published as No. 1 of the Memoirs of the Institute of Fundamental Studies, Sri Lanka

World Scientific

NEW JERSEY · LONDON · SINGAPORE · BEIJING · SHANGHAI · HONG KONG · TAIPEI · CHENNAI · TOKYO

Published by

World Scientific Publishing Co. Pte. Ltd.

5 Toh Tuck Link, Singapore 596224

USA office: 27 Warren Street, Suite 401-402, Hackensack, NJ 07601

UK office: 57 Shelton Street, Covent Garden, London WC2H 9HE

British Library Cataloguing-in-Publication Data
A catalogue record for this book is available from the British Library.

PROOFS THAT LIFE IS COSMIC
Acceptance of a New Paradigm

ISBN 978-981-3233-10-2

Contents

Introduction

The theory of cosmic life started off as a hypothesis to be tested in 1974 after the first discovery of complex organic molecules and polymeric dust in interstellar space. The hypothesis soon developed into a multifaceted scientific theory, predictions of which were available to be verified or falsified. Over four decades there have been a multitude of tests and predictions of the theory, all of which have been positive in the direction of vindicating the proposition of life being a cosmic rather than a purely terrestrial phenomenon. A paradigm shift of enormous magnitude and far-reaching significance lies round the corner.

1. History of Publication

When *Proofs that Life is Cosmic* was first put together in 1982, I was in the midst of planning an international conference on "Frontiers of Science" to held in Sri Lanka under the auspices of the Institute of Fundamental Studies. I was the founding Director of this Institute (IFS) and in view of my own research interests I had arranged for *Proofs that Life is Cosmic* to be one of the main topics of discussion at the meeting.

By this time Fred Hoyle and I had already amassed a significant body of evidence favouring this point of view and published a large number of scientific papers in peer-reviewed journals. The participants of the meeting in Colombo included both adversaries and supporters of this idea, including Gustaf Arrhenius (grandson of Svante Arrhenius), Cyril Ponnamperuma, Asoka Mendis, Hans D. Pflug, Bart Nagy among others, who each brought to bear arguments and counter-arguments that could be objectively assessed. Even if at this stage evidence in favour of this thesis was considerable, such opposition as prevailed was largely polemical.

At the end of 1982, I negotiated with President J.R. Jayawardene, then President of Sri Lanka (as well as Chair of the Board of Management of the Institute of Fundamental Studies) to have *Proofs that Life is Cosmic* printed by the Government Press and issued as the first publication of the fledgling Institute of Fundamental Studies. A few months later, a limited number of copies was circulated in the Cardiff Blue Preprint series of the Department of Applied Mathematics and Astronomy, University College Cardiff.

Since 1982, a veritable raft of evidence from Astronomy, Space probes, Geology and Biology has added strength to the case for life as a cosmic phenomenon. The new evidence even over the past decade, has become absolutely compelling. A major paradigm shift with far-reaching implications appears to be just round the corner, impeded only by longstanding prejudice nurtured over centuries, combined with professional rivalries and clashes of interest. This is the motivation of reissuing *Proofs that Life is Cosmic* in exactly its original form. In this introductory chapter, I discuss a few of the more crucial updates to the data and arguments that have appeared since 1982.

2.   Improbability of the Origin of Life

The success of the famous Miller–Urey experiment of 1958/59 showing how organic molecules could be synthesised from mixtures of inorganic gases under laboratory conditions led to the conviction that it was only a matter of time before the next steps from biochemical monomers to life itself could be demonstrated in the laboratory (Miller and Urey, 1959). Despite over half a century of effort, this goal has proved stubbornly elusive (Deamer, 2012).

Starting from an initial small bacterium (typified by *Mycoplasma genitalium*) with ~500 genes, life has apparently evolved over a 4 billion year timescale to produce the entire spectrum of terrestrial life including mammals and humans with genomes consisting of some 25,000 genes. Hoyle and the present writer dwelt at length on the improbability of obtaining a minimal gene set needed for the emergence of a bacterial genome from random processes. For a set of 500 genes, assuming that about 10 sites per gene need to be correctly filled with one of a set of 20 amino acids, the probability turns out to be $\sim 10^{-6500}$ (Wickramasinghe, 2011, 2015).

If one accepts this calculation showing a grotesquely small *a priori* probability for the transition of non-life to life (see *Proofs that Life is Cosmic*), it would appear that only two options remain open. The origin of life on Earth was an extremely improbable event that somehow occurred (because we are here!) but will effectively not be reproduced again either here or elsewhere. A vastly bigger system of truly cosmological proportions as well as an immeasurably longer timescale were involved in an initial origination event. How big or old that system needs to be is still a matter for debate. Arguments by Abel and Trevors (2006) and Abel (2009) suggest that within the framework of Big Bang type cosmologies naturalistic protogene formation still faces almost insuperable difficulties. However, by whatever process life has emerged, this event of origination must be reckoned as unique and the subsequent spread of life throughout the universe assured by the processes of "panspermia".

3. Interstellar Dust, PAH and Biological Molecules

Identifying the composition of dust in interstellar clouds was the major incentive for embarking on the theory of cosmic life (Wickramasinghe, 1967). The interstellar dust absorbs and scatters starlight causing extinction of the light from stars, re-emitting the absorbed radiation in the infrared. An important clue related to dust composition followed from the studies of the extinction of starlight. From my earliest researches it transpired that the paradigm in the 1960s that the dust was largely comprised of water–ice was quickly overturned. Infrared observations showed absorptions due to CH, OH, C-O-C linkages consistent with organic polymers (Wickramasinghe, 1974; Hoyle *et al.*, 1982). The best overall agreement over the entire profile of interstellar extinction from the infrared to the ultraviolet was a mixture of desiccated bacteria, nanobacteria, including biologically derived aromatic molecules (Hoyle and Wickramasinghe, 1991, 2000, 2015).

The distribution of unidentified infrared bands (UIB's) between 3.3 and 22 μm is almost identical in their wavelengths in widely different astronomical sources, more or less irrespective of the ambient conditions (Rauf and Wickramasinghe, 2010). Data on UIB's for a large number of galactic and extragalactic sources have been obtained using the Spitzer Space Telescope and other orbiting telescopes. Figure 1 shows the spectrum of the planetary nebula NGC 7027 with arrows pointing to the strong absorption bands in an aromatic mixture of chemicals (Wickramasinghe, 2015).

Whilst PAHs (Polyaromatic Hydrocarbons), presumed to form inorganically, are the favoured model for the UIBs, no really satisfactory agreement with available astronomical data has thus far been shown possible for abiotic PAHs (Hoyle and Wickramasinghe, 1991; Rauf and Wickramasinghe, 2010). This is a particularly serious problem if we require the set of UIB emitters and the 2175A absorbers as manifest in the ultraviolet extinction curve to be one and the same. The latter

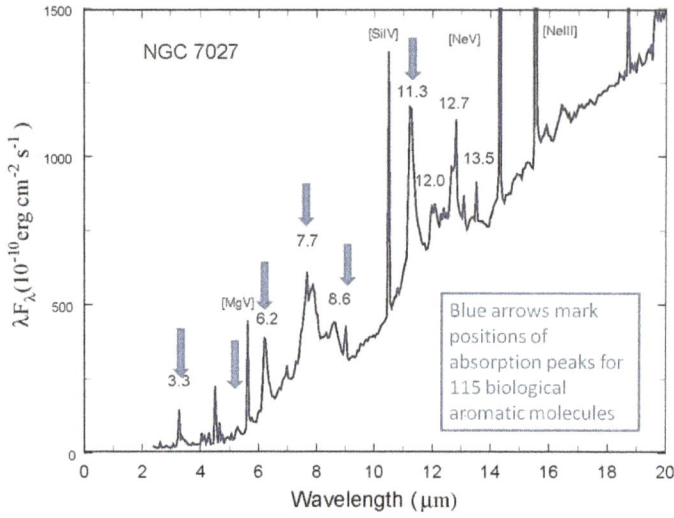

Fig. 1. IR spectrum of planetary nebula NGC7027 compared with main absorptions of 115 biological aromatic molecules.

requirement is necessary because it is the starlight energy absorbed in the ultraviolet band that is being re-emitted as UIBs in the infrared.

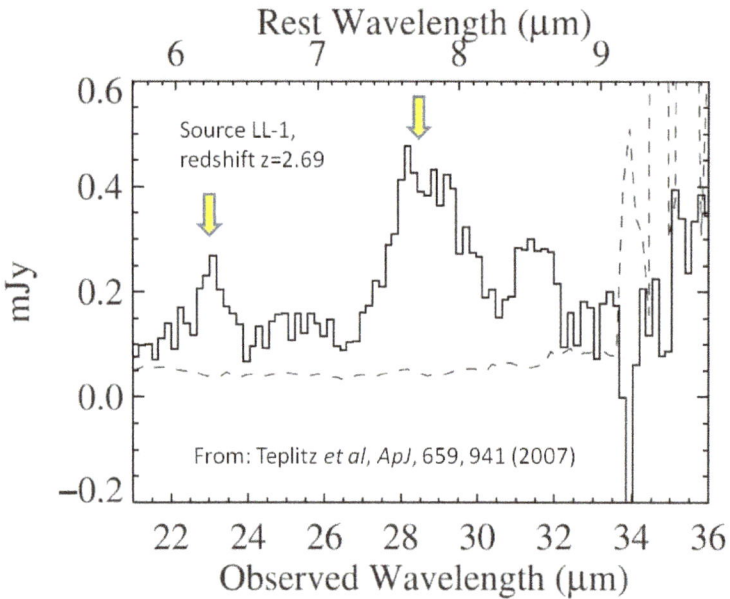

Fig. 2. Redshifted 6.2, 8.7, 11.3 micron bands in the source (Tepletz *et al.*, 2007).

Amongst the most distant galaxies displaying aromatic/biomolecular infrared signatures is a high redshift infrared luminous galaxy at redshift z = 2.69, the spectrum of which is shown in Fig. 2 (Teplitz *et al.*, 2007). This galaxy emitted its light when the Universe was at the tender age of 2 billion years according to standard Big Bang cosmology.

Evidence of PAH-type biodegradation products also shows up in the form of a 2175A spectral signature in recent observations of the most distant galaxies upto redshifts of at least 2.45. We thus have clear evidence that biologically related materials were formed within 2.5 billion years of the presumed Big Bang origin of the Universe (Motta *et al.*, 2002; Noterdaeme *et al.*, 2009). The UV spectral feature for two intermediate redshift galaxies is shown in Fig. 3 compared with our calculated absorption profile for the biological aromatic ensemble that agrees with the IR and the mid UV absorptions in our galaxy (Hoyle and Wickramasinghe, 1991, 2015).

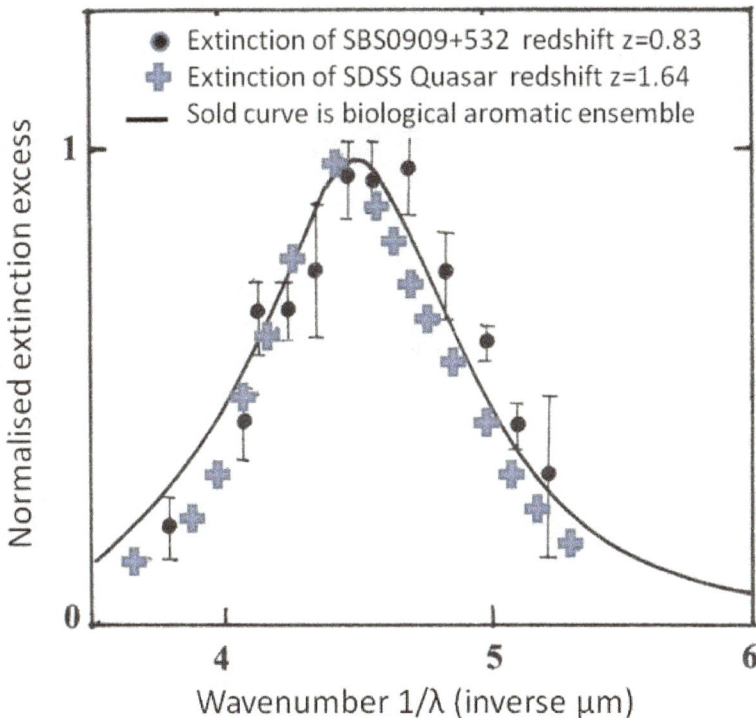

Fig. 3. The 2175A feature in two high redshift galaxies, compared with aromatic biomolecules (Motta *et al.*, 2002; Noterdaeme *et al.*, 2009).

Detections of interstellar organic molecules of ever-increasing complexity including molecules relevant to biology have continued in the last two decades with the deployment of newer and better instruments and telescopes. Infrared, microwave and radio observations are used to detect the presence of such molecules and the current list of positive detections is likely to be constrained only by limitations of techniques.

4.   Comets and Biomaterial

The organic composition of comets and the concept of cometary panspermia were proposed and developed by Hoyle and Wickramasinghe (1981, 1985). The first observational evidence that had a direct bearing on the possible existence of cometary biology (rather than just organics) came with the 1986 perihelion passage of Comet Halley. The mid-infrared spectrum (Fig. 4) of the comet's dust coma following a major outburst on March 31, 1986 showed unambiguous evidence of aromatic–aliphatic linkages (C-H stretching modes) that were consistent with desiccated *E-coli* (Wickramasinghe, D.T. & Allen, 1986; Wickramasinghe *et al.*, 1986).

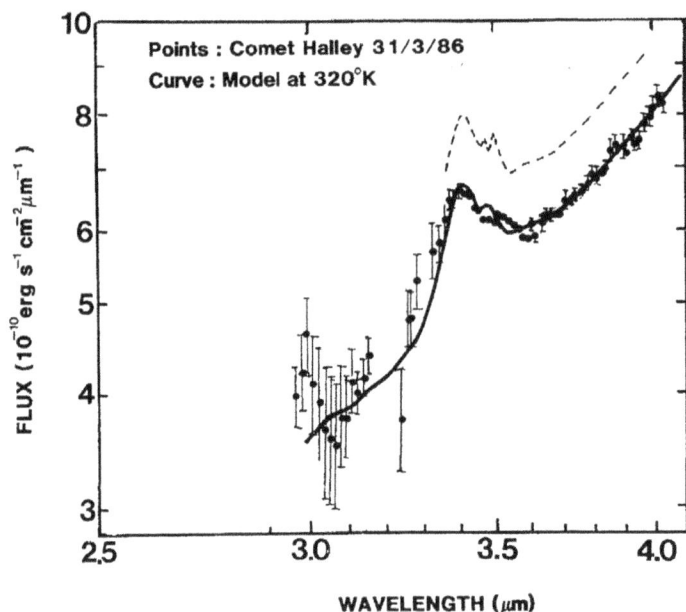

Fig. 4. Emission by dust coma of Comet Halley observed by D.T. Wickramasinghe and D.A. Allen on March 31, 1986 (points) compared with normalized fluxes for desiccated E-coli at an emission temperature of 320K. The solid curve is for unirradiated bacteria; the dashed curve is for X-ray irradiated bacteria (Wickramasinghe, D.T. & Allen, 1986).

Since 1986 many comets have been studied both with ground-based instruments and well as by techniques using spacecraft. All these studies have yielded similar results. Comets, in addition to water ice and silicate minerals have a large component of organic material that could most plausibly be explained in terms of biology and biological processing.

Most recently, the European Space Agency's Rosetta Mission to comet 67P/C-G has yielded the most detailed observations that satisfy consistency checks for biology and our ideas of cometary panspermia. Figure 5 shows the close consistency between the surface properties of the comet and the spectrum of a desiccated bacterial sample.

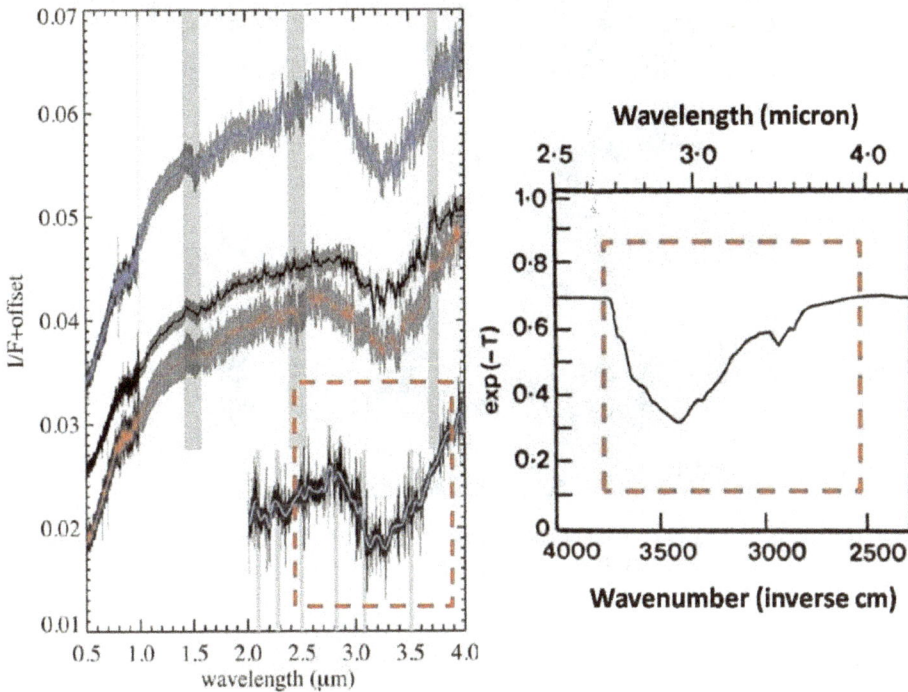

Fig. 5. The surface reflectivity spectra of comet 67P/C-G (left panel) compared with the transmittance curve measured for *E-coli* (right panel) (Capaccione *et al.*, 2015).

Jets of water and organics were observed issuing from ruptures and vents in the frozen surface (Fig. 6), again consistent with biological activity occurring within sub-surface liquid pools (Wallis & Wickramasinghe, 2015). The most recent report of O$_2$ along with evidence for the occurrence of water and organics provides further evidence of such ongoing biological activity (Bieler *et al.*, 2015). Such a mixture of gases cannot be produced under thermodynamic conditions, because organics are readily destroyed in an oxidizing environment. The freezing of an initial mixture of compounds, including O_2, not in thermochemical equilibrium, has been proposed, but there is no evidence to support such a claim. On the other hand the oxygen/ water/organic outflow from the comet can be explained on the basis of subsurface microbiology. Photosynthetic microorganisms operating at the low light levels near the surface at perihelion could produce O_2 along with organics. Many species of fermenting bacteria can also produce ethanol from sugars, so the recent discovery that Comet Lovejoy emits ethyl alcohol amounting to 500 bottles of wine per second may well be an indication that such a microbial process is operating (Biver *et al.*, 2015).

Fig. 6. September, 10, 2014 imaging shows jets of cometary activity along the whole neck of the comet. Credit: ESA/Rosetta/MPS for OSIRIS Team MPS/UPD/LAM/ IAA/SSO/INTA/UPM/ DASP/IDA.

The outbursts of comet 67P/C-G could not result from the volatilisation of the comet's surface, but from gas pressure built up in subsurface (presumed liquid) domains. The contents of a pressurised 10^6–10^7 tonne lake flooding out with say 10% as gases and entrained particles moving into the coma, could explain the outbursts of dust observed in comets (Wickramasinghe *et al.*, 1996; Wickramasinghe *et al.*, 2015). Although CO, CO_2 or CH_4 may well be constituents of the driver gas at the base of ruptured vents, most of this material would recondense on grains close to the surface while liquid water would form an ice crust due to sublimation cooling. The CO production (observed in comet Hale-Bopp) would most naturally be explained as a photodissociation product of volatile and fragile biochemicals that contain weakly bonded CO groups which re-evaporate from superheated (smaller) grains.

5. Capture of Comet Dust in the Stratosphere

An obvious place to find fragile particles from comets is the Earth's stratosphere. Cometary meteoroids and cosmic dust particles are known to arrive at the Earth at a more or less steady rate, averaging $\sim 10^2$ tonne/day. Although much of this incoming material burns up as meteors, a significant fraction survives entry. Organic grains of micron sizes, arriving as clumps and dispersing in the high stratosphere, would be slowed down gently and would not be destructively heated. The Earth's atmosphere could thus serve as an ideal collector of organic cometary dust.

Techniques for stratospheric collection of cometary dust must of necessity involve procedures for either sifting out terrestrial contamination or for excluding such contamination altogether. Stratospheric dust collections have been carried out from as early as the mid-1960s (Gregory and Monteith, 1967). Balloons and rockets reaching heights well above 50 km were deployed and consistently brought back algae, bacteria and bacterial spores. Although some of the microorganisms thus collected were claimed to exhibit unusual properties, such as pigmentation and radiation resistance, their possible extraterrestrial origin remained in doubt. No DNA sequencing procedure was available at the time to ascertain any significant deviations there might have been from related terrestrial species. Moreover, both the methods of collection and the laboratory techniques available in the 1960s left open a high chance of contamination.

In 2001, the Indian Space Research Organisation (ISRO) collaborated with us in Cardiff UK to launch a balloon into the stratosphere carrying devices to collect stratospheric air under aseptic conditions (Harris *et al.*, 2002). The procedure involved the use of cryogenically cooled stainless steel cylinders that were evacuated and fitted with valves that could be opened when they reached a predetermined altitude. Large quantities of stratospheric air containing dispersed particulate matter were thus collected from heights upto 41km and brought back for analysis.

The ultra-high pressure stratospheric air contained within the cylinders was carefully released and passed through a system including millipore membrane filters. Upon such filters stratospheric aerosols were collected, extreme care being taken at every stage to avoid contamination. The particles that were collected fell into two broad classes: (a) mineral grain aggregates, very similar to Brownlee particles, but somewhat smaller; (b) fluffy carbonaceous aggregates resembling clumps of bacteria (see Fig. 7). Typical dimensions were about 10μm.

The cometary origin of such particles is very strongly indicated, the altitude of 41 km being too high for lofting 10-μm sized clumps of solid material from the Earth's surface. In addition to structures such as shown in Fig. 7, which can be tentatively identified with degraded bacteria, the stratospheric samples also revealed evidence of similar sized bacterial clumps that could not be cultured, but were nevertheless determined to be biological by the use of a fluorescent dye (carbocyanine). The uptake of the dye revealed the presence of viable but not culturable living cells in the clumps.

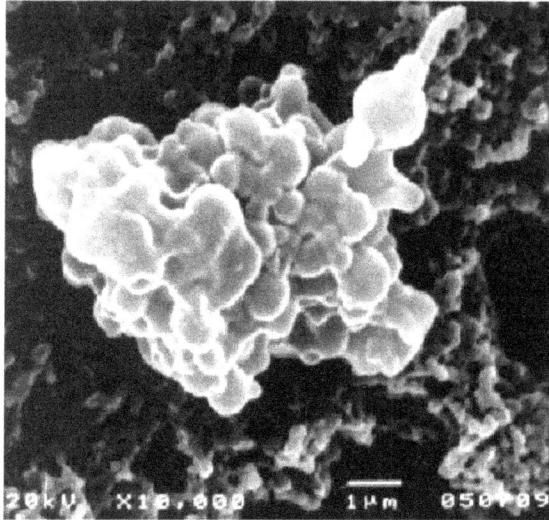

Fig. 7. A carbonaceous stratospheric particle from 41 km resembling a clump cocci and a rod bacterium (Harris *et al.*, 2002).

This data was the last direct confirmation of the cosmic life theory that Fred Hoyle was able to see before his death in 2001.

In a separate series of experiments, a few microbial species were cultured from the same stratospheric air samples by Wainwright *et al.*, (2003), and there is tentative evidence that these may have come from comets. A later balloon flight in 2008 collected more stratospheric material, and analysis by S. Shivaji and his colleagues (Shivaji *et al.*, 2009) yielded cultures of three hitherto unknown microbial species which were highly resistant to ultraviolet light. One of the new species was named *Janibacter hoylei*, in honour of Fred Hoyle.

From 2013 onward, Milton Wainwright has continued to deploy balloons to collect samples from the stratosphere upto heights of approximately 30km (Wainwright *et al.*, 2015). The recovered particulate material from such flights included hitherto unidentified biological entities as well as a fragment of a diatom frustule (see Fig. 8), all of which arguably had a space origin and were not terrestrial contaminants.

Perhaps the most dramatic case of recovering cometary microorganisms was the 2015 report by a team of Russian cosmonauts who claimed the recovery of a variety of marine diatom species from the outside of the International Space Station that orbits the Earth at a height of 400km (Tsygankov *et al.*, 2015). No claim of terrestrial contamination appears to be tenable in this instance and here is perhaps the clearest direct proof of cometary organisms entering the Earth's vicinity.

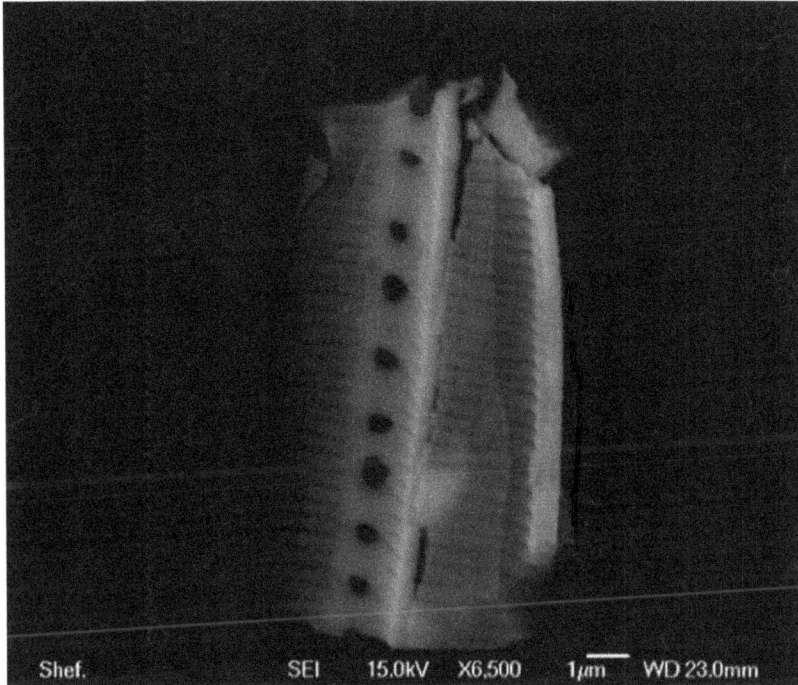

Fig. 8. Fragment of diatom frustule collected from 27 km altitude in the stratosphere by Milton Wainwright's team.

6. Microfossils in Meteorites

The topic of microfossils in carbonaceous chondrites has sparked bitter controversy ever since it was first introduced in the mid-1960s (Claus *et al.*, 1963). Since carbonaceous chondrites are generally believed to be derived from comets, the discovery of fossilised life forms in comets would provide strong *prima facie* evidence in support of the theory of life in comets and cometary panspermia. However, some contamination was discovered in the original samples, and claims that all micro structures (organised elements) discovered in meteorites were artifacts or contaminants led to a general rejection of meteoritic microfossil identifications in the 1960s.

The situation remained so until early in 1980 when H.D. Pflug found a similar profusion of "organised elements" in ultrathin sections prepared from the Murchison meteorite, a carbonaceous chondrite that fell on 28 September 1969 in Australia (Pflug, 1984). The contaminant free experimental methods adopted by Pflug appeared to be beyond reproach. Thin sections of the meteorite were placed on membrane filters and hydroflouric acid was used to dissolve the bulk of the minerals present and the residue examined in an electron microscope. The morphologies of the residues were strikingly characteristic of particular types of fossil microbes, and furthermore laser ion probe analysis yielded elemental and molecular compositions within these structures that were also consistent with life. These studies made it very difficult to reject the fossil identification.

More recent work by Richard Hoover (2005) provided even more striking evidence of microbial fossils in the Murchison meteorite as shown in Fig. 9.

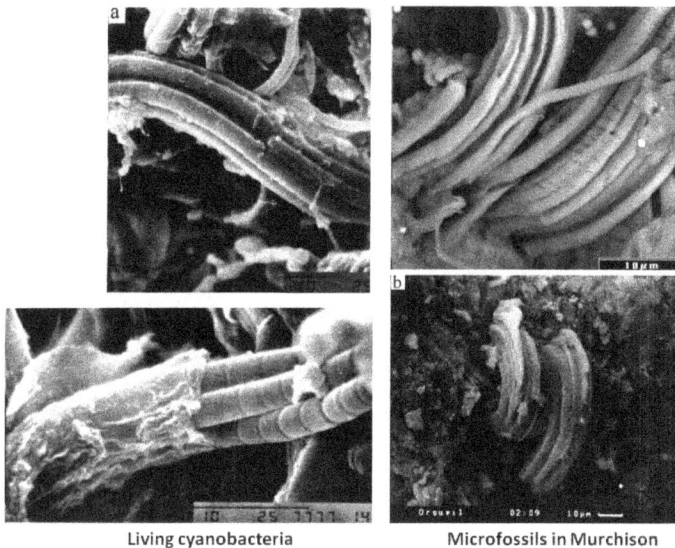

Living cyanobacteria   Microfossils in Murchison

Fig. 9. Structures in the Murchison meteorite compared with living cyanobacteria (Hoover, 2005).

A development of particular significance relates to a recent meteorite fall in Sri Lanka. Minutes after a huge fireball was seen by a large number of people in the skies over Sri Lanka on 29 December 2012, a large meteorite disintegrated and fell in the village of Araganwila, which is located a few miles away from the historic ancient city of Polonnaruwa. At the time of entry into the Earth's atmosphere the parent body of this Polonnaruwa meteorite would have had most of its interior porous volume filled with water, volatile organics and possibly viable living cells. Fragments from a freshly cleaved interior surface of the Polonnaruwa meteorite were mounted on aluminium stubs and examined by Jamie Wallis under an environmental scanning electron microscope (Wickramasinghe *et al.*, 2013).

Images of the sample at low magnification displayed a wide range of structures that were distributed and enmeshed within a fine-grained matrix, examples of which are shown in Fig 10. They include fresh water and seawater diatoms and an extinct microbial fossil (top left) known as an acritarch. Some of these structures are deeply ingrained in the rock matrix and the range of species we found cannot be reasonably

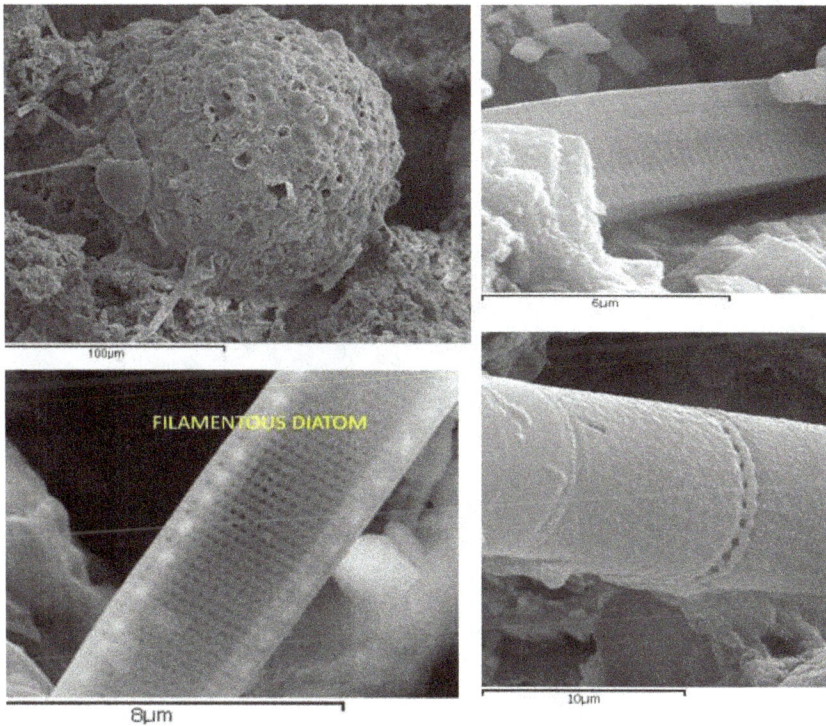

Fig. 10. Fossilised acritarch (top left) and diatoms in the Polonnaruwa meteorite (Wickramasinghe *et al.*, 2013).

explained on the basis of contamination. Although the space origin of these rocks has been called to question by critics, studies of the distribution of oxygen isotopes leave little doubt of their cosmic origin (Wickramasinghe *et al.*, 2013).

We conclude therefore that the identification of fossilised diatoms in the Polonnaruwa meteorite is established beyond reasonable doubt although the endorsement of this conclusion by the wider meteoritic community has been slow to come.

7. Evidence from Geology

Three decades ago, the earliest evidence for microbial life in the geological record was thought to be in the form of cyanobacteria-like fossils dating back to 3.5 Ga ago. From the time of formation of a stable crust on the Earth 4.3 Ga ago following an episode of violent impacts with comets (the Hadean Epoch), there seemed to be available a 800 million years timespan during which the canonical Haldane-Oparin primordial soup may have developed. Very recent discoveries, however, have shown that this time interval has been effectively closed. Detrital zircons older than 4.1Ga, discovered in rocks belonging to a geological outcrop in the Jack Hills region of Western Australia, have been found to contain micron-sized graphite spheres with an isotopic signature of biogenic carbon (Bell *et al.*, 2015) [Fig. 11].

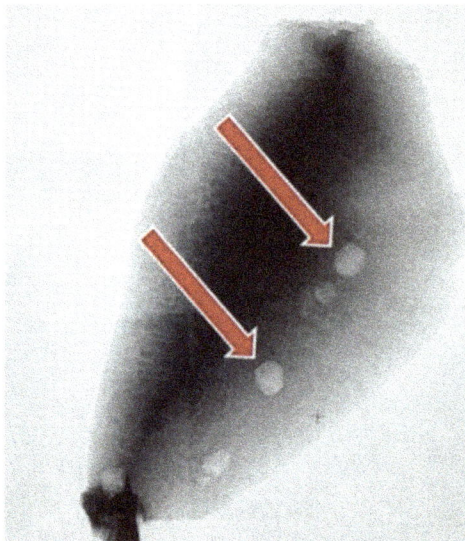

Fig. 11. The ^{12}C enriched micron-sized graphite spheres showing evidence of degraded bacteria within zircon crystals (Bell *et al.*, 2015).

The ^{12}C-enrichment found within these inclusions may thus be taken as unequivocal evidence for the existence of microbial life on Earth before 4.1 Ga, during the epoch of comet and asteroid impacts. The requirement now, on the basis of orthodox thinking, is that an essentially instantaneous transformation of nonliving organic matter to bacterial life is demanded, a fact that strains credibility of Earth-bound abiogenesis beyond the limit. A far more plausible possibility is that fully-developed microorganisms arrived at the Earth via impacting comets, and these became carbonized and trapped within condensing mineral grain conglomerates. The scientific consensus, whilst moving away from an organic soup generated *in situ* on Earth to one supplied from space still clings tenaciously to a preferred origin of life on the Earth.

8. Evidence from DNA Sequencing

After the human genome was fully sequenced our cosmic ancestry was essentially laid bare (Venter *et al.*, 2001; Villareal, 2004; Wickramasinghe, 2012). The discovery that only 25,000 genes exist to code for all our proteins and enzymes left the vast majority of our seemingly redundant DNA to be comprised of viral derivatives. There is now little doubt that much of our genetic inheritance may be comprised of DNA actually delivered by viruses. They show up as LINEs (Long Interspersed Nuclear Elements) (21%), and SINEs (Short Interspersed Nucluear Elements) (13%) which are retroviral derived and controlled, and HERVs (Human Endogenous Retroviruses) and LTRs (Long Terminal Repeats) (9%).

The viewpoint developed in *Proofs that Life is Cosmic* in 1982 that viral genes of external origin are continually added to genomes of evolving life-forms has become a fact that is now difficult to deny. Therefore, the role of neo-Darwinian evolution occupies a less important role as a kind of fine-tuning, rather than as the main driving force. Major evolutionary traits in the development of complex life are all externally derived, and evolution is essentially driven from outside. If this is so the overall impression will be of a pre-programming in the higher levels of development in biology. The mechanism is that the relevant viral genes that were transported to Earth had evolved over vast timescales and in innumerable locations and suddenly came to be expressed locally on Earth. The evolution of the eye may be seen as one example of this type, and even some highly complex, and less definable manifestations of gene expression in our own immediate line of descent in hominid evolution bear the signs of "pre-programming".

If a single discovery is to serve as a watershed in the journey to validating the theory of cometary panspermia it is a recent study that concerns the Octopus. The octopus genome has recently been sequenced and revealed a staggering level of complexity with 33,000 protein-coding genes more than in a human being (Albertin *et al.*, 2015). Octopus belongs to the coleoid subclass of molluscs that have an evolutionary history that stretches back over 500 million years. The genetic divergence of octopus from its ancestral coleoid subclass is on a truly astronomical scale. Its large brain and sophisticated nervous system, camera-like eyes, flexible bodies, ability to switch colour are just a few of the amazing features that appear suddenly on the scene. The transformative genes leading from squid to octopus are not to be found in any preexisting life form at the time — in a sense they seem "borrowed" from a far distant "future" in terms of terrestrial evolution, or more realistically from the cosmos at large.

References

Abel, D.L. and Trevors, J.T. (2006) *Phys. Life Rev.* **3**, 211.

Abel, D.L. (2009). *Theor. Biol. Med. Model* **6**(1), 27. Open access at www.tbiomed.com/content/6/1/27.

Albertin, C.B., Simakov, O., Mitros, T. *et al.* (2015). *Nature* **524**, 220–224.

Bell, E.A., Boehnke, P., Harrison, T. *et al.* (2015). Potentially biogenic carbon preserved in a 4.1 billion-year-old zircon, *PNAS*, www.pnas.org/cgi/doi/10.1073/pnas.1517557112

Bieler, K. *et al.* (2015). Abundant molecular oxygen in the coma of Comet 67P/C-G, *Nature* **526**, 678.

Biver, N. *et al.* (2015). Ethyl alcohol and sugar in comet C/2014Q2 (Lovejoy) *Sci.Adv.1*, e1500863 23 October, 2015.

Claus, G., Nagy, B. and Europa, D.L. (1963). *Ann. New York Acad. Sci.* **108**, 580.

Deamer, D. (2012). *First Life.* (University of California Press).

Gregory, P.H. and Montieth, J.L. (eds.) (1967). Airborne microbes in *Symp. Soc. General Microbiology,* Vol. 17 (Cambridge University Press).

Harris, M. J., Wickramasinghe, N.C., Lloyd, D. *et al.* (2002). *Proc. SPIE* **4495**, 192.

Hoover, R.B. (2005) In: eds. Hoover, R.B., Rozanov, A.Y. and Paepe, R.R., *Perspectives in Astrobiology* **366**, 43. (IOS Press: Amsterdam)

Hoyle, F. and Wickramasinghe, N.C. (1981). In: ed. *Comets and the Origin of Life.* Ponnamperuma, C., Reidel, D.: Dordrecht, 227 pp.

Hoyle, F. and Wickramasinghe, N.C. (1985). *Living Comets.* (Univ. College, Cardiff Press).

Hoyle, F. and Wickramasinghe, N.C. (1991). *The Theory of Cosmic Grains.* (Kluwer, Dordrecht).

Hoyle, F. and Wickramasinghe, N.C. (2000). *Astronomical Origins of Life: Steps towards Panspermia.* (Kluwer Academic Press: Dordrecht).

Hoyle, F., Olavesen, A.H., and Wickramasinghe, N.C. (1978). Identification of interstellar polysaccharides and related hydrocarbons, *Nature* **271**, 229–231.

Miller, S.L. and Urey, H.C. (1959). Organic compound synthesis on primitive Earth, *Science* **130**, 245.

Motta, V., Mediavilla, E., Munoz, J.A. *et al.* (2002). Detection of the 2175A Extinction Feature at z=0.83, *Astrophys J.* **574**, 719.

Noterdaeme, P. *et al.* (2009). Diffuse molecular gas at high redshift, *Astron. Astrophys* **503**, 765.

Pflug, H.D. (1984). In *Fundamental Studies and the Future of Science,* ed. Wickramasinghe C. (University College, Cardiff Press).

Rauf, K. and Wickramasinghe, C. (2010). *Int. J. Astrobiol.* **9** (1), 29–34.

Shivaji, S. *et al.* (2009). *Int J. Syst. Evolut. Biol.* **59**, 2977–2986.

Teplitz, H.I. *et al.* (2007). Measuring PAH emission in ultradeep Spitzer IRS spectroscopy of high redshift IR-luminous galaxies, *Astrophys. J.* **659**, 941.

Tsygankov, O.S., Grebennikova, T.V., Deshevaya, E.A. *et al.* (2015). On the outer surface of the international space station and detection of microbiological objects in space experiment test, *The Space Engin. Technol.* **1** (8).

Venter J.C.J., Adams, M.D., Myers, E.W., Li, P.W., Mural, R.J., *et al.* (2001) The sequence of the human genome, *Science* **291**, 1304–1351.

Villarreal, L.P. (2004). Can viruses make us human? *Proc American Phil. Soc.* **148**, 296–323.

Wainwright, M., Rose, C.E., Baker, A. J., Bristow, K.J., Wickramasinghe, N.C. (2013). Isolation of a diatom frustule fragment from the lower stratosphere (22-27Km)- Evidence for a cosmic origin, *J. Cosmol.* **22**, 1063.

Wainwright, M., Wickramasinghe, N.C., Narlikar, J.V. and Rajaratnam, P. (2003). *FEMS Microbiol. Lett.* **218**, 161.

Wallis, M.K., Wickramasinghe, N.C. (2015). Rosetta images of comet 67P/Churyumov-Gerasimenko: Inferences from its terrain and structure, *Astrobiol. Outreach* **3**, 12.

Wickramasinghe, C. (2011). Viva panspermia, *The Observatory* **131**, 130–134.

Wickramasinghe, N.C. (2012). DNA sequencing and predictions of the cosmic theory of life, *Astrophys. Sp. Sci.* **343**, 1–5.

Wickramasinghe, N.C., Wallis, J. and Wallis, D.H. (2013). *Mod. Phys. Lett. A* **28**, No. 14.

Wickramasinghe, C. (2015). *The Search for Our Cosmic Ancestry.* (World Scientific Publ., Singapore).

Wickramasinghe, D.T. and Allen, D.A. (1986). Discovery of organic grains in Comet Halley, *Nature* **323**, 44–46.

Wickramasinghe, D.T., Hoyle, F., Wickramasinghe, N.C. and Allen, D.A. (1986). *Earth, Moon and Planets* **36**, 295.

Wickramasinghe, N.C. (1967). *Interstellar Grains.* (Chapman & Hall: London).

Wickramasinghe, N.C. (1974). Formaldehyde polymers in interstellar space, *Nature* **252**, 462–463.

Wickramasinghe, N.C., Hoyle, F. and Lloyd, D. (1996). Eruptions of comet Hale-Bopp at 6.5 AU., *Astrophys. Sp. Sci.* **240**, 161–165.

Wickramasinghe, N.C., Wainwright, M., Smith, W.E., Tokoro, G., Al Mufti, S. and Wallis, M.K. (2015). Rosetta studies of comet 67P/ Churyumov–Gerasimenko: Prospects for establishing cometary biology, *J. Astrobiol. Outreach* **3**, 1.

Memoirs of the Institute of Fundamental Studies, Sri Lanka, No. 1

Prof. Sir Fred Hoyle

and

Prof. Chandra Wickramasinghe

PROOFS THAT LIFE IS COSMIC

December 1982

1–A 66857 (12/82)

PREFACE

THE arguments for why life is cosmic are many and diverse. In the past we have published our work on these matters in different places, as papers or preprints and as books, choosing the mode of presentation more or less to fit traditional practices. This scattering of the arguments has by now made it awkward for others to see how all the pieces of the jig-saw fit together, to a point where we have felt it necessary to put together the whole thing between the same covers.

Nowadays few of us have the wish to follow extensive arguments in detail, the peaceful days of Trollope's Barchester Towers being long since gone. It has therefore seemed best to break-up the whole into small sections, as in the entries of a catalogue, entries which can be read singly in isolation from the rest. Besides being an advantage to the reader, such a procedure avoids the temptation to support one argument with another too much, a procedure that can easily become like drunks supporting each other at a party. To compile our catalogue we have divided the entire subject into nine principal sections, giving them the letters A, B, C, D, E, I, M, O and P, with the following designations :–

A	=	Atmosphere (terrestrial)
B	=	Bacteria and other microorganisms
C	=	Comets
D	=	Diseases
E	=	Evolution (biological)
I	=	Interstellar grains
M	=	Meteorites
O	=	Origin of life
P	=	Planets

Within each of these main sections, subsections are classified numerically, as A1, A2, ... for instance. Tables and figures are named according to the subsection in which they occur, Table A 1.1, Table A 1.2, Figure M 1.1, Figure M 1.2, ...and so on. A similar system is used for the pagination. The subsections are cross-referenced wherever an explicit mention of the inter-relation of the arguments seemed unavoidable. References are given in the text as they occur.

Fred Hoyle

Chandra Wickramasinghe

CONTENTS

A

Atmosphere (terrestrial)

A 1 The Safe Entry of Microorganisms into the Earth's Atmosphere

Particles are least heated when they enter the atmosphere at a glancing angle, in the fashion of returning astronauts The present authors obtained the formula :

$$a = 24 \frac{\sigma\ T^4}{v^3} \sqrt{HR} \qquad\qquad A\ 1.1$$

for a spherical particle (Diseases from Space, page 171, J. M. Dent 1979), where a is the particle radius, $\sigma = 5.669 \times 10^{-5}$ is the Stefan-Boltzmann constant, T in degrees K is the temperature of a flash heating of the particle which lasts for a few seconds, v is the entry speed, H the atmospheric scale height at the altitude where the particle is effectively decelerated, and R is the radius of the Earth (which appears because of the glancing angle of entry).

Microorganisms can probably withstand a short flash heating for values of T up to 500 K. Putting this temperature value in A1.1, together with R=6378 km, and setting H=20 km for an altitude of 130 to 140 km above the Earth's surface (where small particles are

1

stopped) the maximum radius of safe entry for a microorganisms can be calculated if v is known. The largest microorganisms for which safe entry is permitted occurs when v is least. For a particle of cometary origin v is least when the following conditions are satisfied :

(i)   The plane of the comet's orbit around the Sun is the same as the plane of the Earth's orbit.

(ii) The perihelion distance of the comet is the same as the radius of the Earth's orbit.

(iii) The comet is in direct motion around the Sun.

When these conditions are satisfied v is about 10 km per second, and equation A1.1 gives a ≈ 30 μm. (Note that it is unnecessary to increase v because of the attraction of terrestrial gravity. The reason is given in the above reference.).

The coefficient 24 appearing in equation A1.1 is specific to a particle of spherical shape. For a rod-shaped particle with 2a the rod diameter the coefficient is changed to 16, giving a maximum rod diameter of about 40 μm. In this case there is no restriction on the length of the rod. This upper limit of ~ 40 μm is calculated for a blunt-nosed particle. Organisms that happned to have favourable aerodynamic shapes could be several times larger still, say with diameters of ~ 100 μm.

When v has its least value there is evidently no restriction on the entry of viruses, bacteria and typical eukaryotic cells. Even whole colonies of bacteria could be admitted, as well as microfungi and most protozoa. If one is concerned with the long-term population of the Earth by microorganisms of extraterrestrial origin, it is the case where v is least that is most relevant (because in a long-term situation one can wait for the most favourable case to arise) but if one is concerned with a short-term pathogenic attack by a microorganisms then v should be given a typical value of 30 to 40 km per second, in which case A1.1 leads to a maximum diameter for safe entry of about 1 μm, in good agreement with the diameters of micrococci. For rod-shaped bacteria 2/3 μm is a typical diameter for safe entry. For rod-shaped bacteria there is again no restriction on their lengths.

A 2. The Amount of Particles Entering the Earth's Atmosphere

The amount of small particles entering the Earth's atmosphere from interplanetary space is usually estimated to be about 1,000 tons per year. The mean geocentric velocity of larger particles, visual or radar meteors, is about 40 km per sec. If the small particles also have this mean geocentric speed, an influx of 1,000 tons per year would imply an interplanetary density of ~ 10^{-23} g cm^3, a reasonable value.

The most direct evidence of the composition of small particles entering the atmosphere was obtained by D. E. Brownlee (*Protostars and Planets*. ed. T. Gehrels, Univ. of Arizona, 1978), who found a CI carbonaceous chondritic composition for particles with overall diameters of ~ 10 µm which were collected in U-2 flights and then subjected to laboratory examination. The CI composition implied that the particles were of cometary origin.

If one combines Brownlee's result with the discovery of microoganisms in the Murchison meteorite by H. D. Pflug (private communication) a considerable fraction of the ~ 5 per cent concentration of carbon found by Brownlee would seem to have had a biological origin. This conclusion is supported by the remarkable similarity between a sticklike object present in one of Brownlee's particles and bacterium found in Murchison by Pflug.

If one says that ~ 1 per cent of the mass of the particles is carbon of biological origin, the mass of microorganisms entering the Earth's atmosphere would be of the order of **10 tons per year**. While microorganisms in the larger particles would be cauterised by heat, those in smaller particles would survive, and since the latter probably contribute a considerable fraction of the total, the surviving microorganisms could well have a combined mass **of the order of 10 tons** per year, giving an entry of viable bacteria of ~ 10^{20} per year, and a possible entry of viruses some two to three powers of ten greater still. If such microorganisms survive descent through the atmosphere, the **vertical** incidence at ground level would be ~ 10^5 bacteria $m^{-2}yr^{-1}$, and perhaps ~ 10^7 viruses $m^{-2} yr^{-1}$, sufficient to be detectable especially if incident microorganisms are pathogenic to plants or animals, in which case there would be a large multiplication of numbers in the bodies of their victims.

Reference to equation A1.1 shows that particles with radii above 1 mm entering the atmosphere at a typical speed of 40 km per sec are flash-heated to their evaporation temperatures. These are the visual meteors. G A. Harvey (Astrophys. J., 1977, 217, 688) observed a strong spectral feature at 3100 A in the radiation emitted by a visual meteor of the α Capricornid stream. The feature was attributed to hydroxyl (OH) derived from dissociated water. While this does not prove the presence of water in smaller particles, it is indicative that water may well be present.

A 3. An Explicit Detection of Bacteria in the High Terrestrial Atmosphere

In *Mikrobiologiya*, 1979, 48, 1066, S. V. Lysenko describes an experiment designed to sample the atmosphere for micro-organisms at heights above the stratopause, i.e. at altitudes above ~ 50 km. Rockets fired into the high atmosphere expelled the detection equipment attached to a parachute. Film was exposed over various height ranges, with

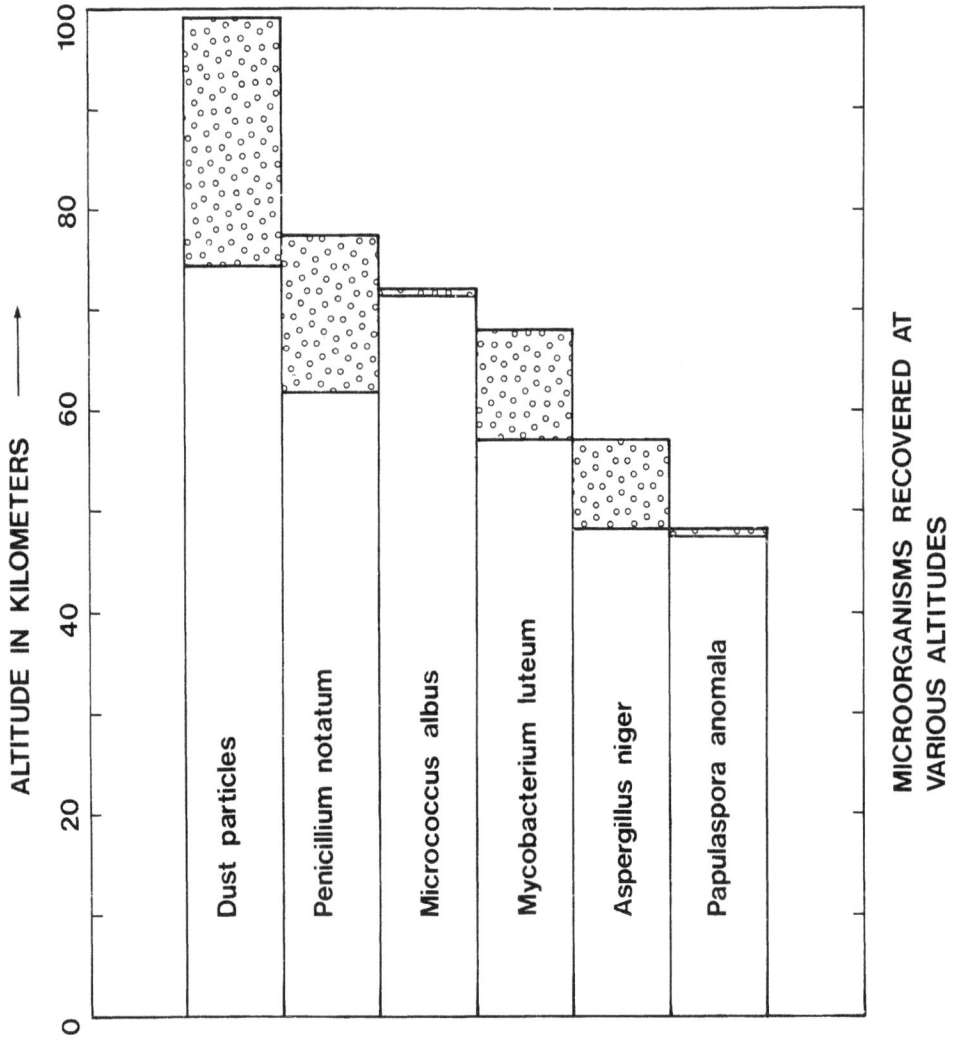

Fig. A 3.1.–Shaded areas show altitude ranges of recovery for various microorganisms in the atmosphere according to S.V. Lysenko (1979)

particles collected on the film being sealed as the equipment descended out of the height range in question. Recovered film was then examined in the laboratory for viable microorganisms.

After three such flights, some thirty cultures were grown of bacteria obtained from altitutdes ranging from about 50 km up to 75 km. The several cultures identified by Lysenko are shown in Fig. A3.1 as a plot of species against altitude of recovery. The cultures were said to be very heavily pigmented, a circumstance attributed to the need of the organisms to withstand unshielded solar ultra-violet radiation at heights above the ozone layer.

Much of Lysenko's paper is concerned with a discussion of why the author considers there could have been no contamination to vitiate his results. The author attributes the recovered bacteria to a large desert storm in which **microorganisms** were carried upward through the stratosphere, a supposition which seems unlikely in view of the great vertical stability of air in the ozone layer, a stability conferred by a strong termperature inversion. Other than an origin from space, the only suggestion consistent with atmospheric physics we have heard is that surface bacteria might have been impelled upward to great heights by a volcanic eruption. If so, there must have been a long persistence of viability, requiring resistance to solar ultraviolet for months or even years.

Although there are vertical air movements in the mesosphere above 50 km to hold up particles that are small enough, gravity has a significant downward pull on particles with the sizes of bacteria (see entry A4), particularly in the thin air above 50 km. This makes it hard to see how bacteria could stay in the mesosphere for more than a few days. (For a typical bacterial size, gravity generates a downward speed of ~ 3 cm per sec in still air at an altitude of 60 km, and a downward speed of ~ 10 cm per sec at an altitude of 70 km.) This makes for a difficulty with the volcanic explanation, unless it happened by a fluke that the time of the experiment was more or less contemporaneous with a large volcanic eruption. As far as we are aware this was not so.

A 4. The Time of Fall of Bacteria through Still Air in the Stratosphere

The time T of fall of a particle through still air in the stratosphere, assuming the only forces to be gravity and the viscous drag of the air, is given by–

$$T = \int_{z_1}^{z_2} \frac{dz}{w(z)}. \qquad\qquad \text{A 4.1}$$

where $w(z)$ is the downward speed of the particle at height z, and the stratosphere is taken to extend from height z_1 up to height z_2. Values of $w(z)$ for particles of various radii in μm are given in Figure A 4.1. Although this figure refers explicitly to spheres of density 1 g cm^{-3}, it can be used for rod-shape bacteria as well as micrococci, with the rod-radius interpreted as ' radius '–the length of the rod is **largely irrelevant**.

5

The time of fall from the base of the stratosphere to ground-level is much less than A4.1, because of vertical air movements in the troposphere, i.e. from height z_1 downwards. Likewise the time of fall through the mesosphere to height z_2 is much less than A4.1, which therefore gives (under the initially stated assumptions) an estimate of the time required for a bacterium incident from interplanetary space to reach ground-level. Inspection of Figure A4.1 shows that for a specified value of the particle radius the time T is essentially a function of z_1. The height z_2 of the top of the stratosphere, the stratopause, enters the calculation of T only weakly.

Fig. A 4.1.–Falling speed of spherical particles of various radii r as a function of height z (Kasten, F., J. Appl. Met., 7, 944, 1968).

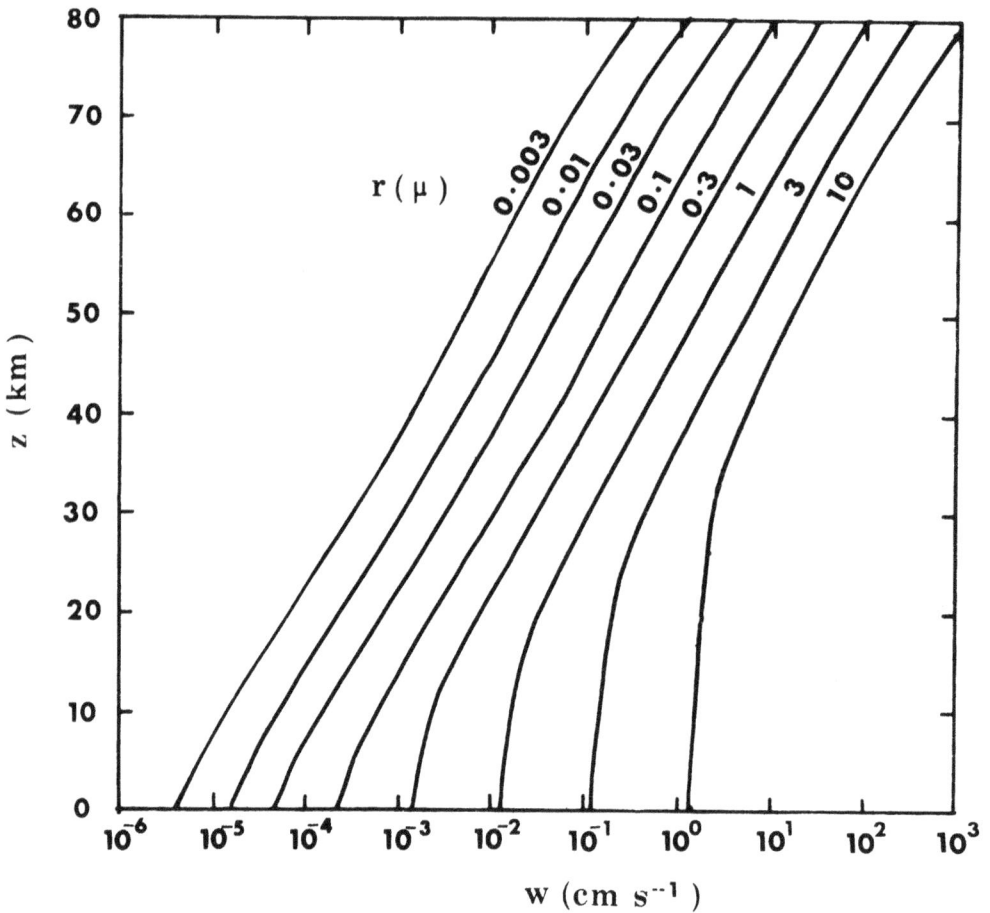

6

The value of z_1 is latitude dependent, being about 17 km at the equator and ～ 10 km in middle latitudes. The generally decreasing value of z_1 with increasing latitude causes T to increase with latitude, particularly as the main contribution to T comes from the lower values of the height coordinate z, because w(z) decreases with z. Thus the time of fall of a bacterium at the equator is about 3 years for a particle radius of 0.3 μm and is about 5 years at higher temperate latitudes. These estimates have relevance to the delay at high-temperate latitudes of about two years in the outbreak of the Black Death, the delay after the disease appeared in 1347 A.D. in lower latitudes (entry D5).

Since ozone absorbs solar ultraviolet light there is in effect a heat source in the stratosphere. Irregularities in the ozone distribution can therefore produce irregularities in the temperature distribution, which may locally destroy the overall temperature inversion of the stratosphere, permitting local vertical movements of the air. Such effects could modify the value of T calculated from A4.1, as also could the presence of localised clouds of volcanic dust. The main caution to be applied to A4.1, however, comes from violent storm conditions in the lower troposphere. Storms can generate upward-moving air currents that penetrate some way into the stratosphere before becoming damped-out by the dissipative processes arising from the stratospheric temperature inversion. Thus A4.1 tends to exaggerate the contribution to T of height intervals close to z_1. Some allowance for this effect was included in the time estimates given above.

A 5. The Fall of Virus-Sized Particles through the Stratosphere

Figure A4.1 shows that virus-sized particles (less than 100 nm up to about 300 nm) would mostly fall appreciably more slowly than bacteria through the stratosphere if no other forces than gravity and the viscous drag of the air were involved. However, for particles appreciably smaller than bacteria electrical effects must also be considered. The process is one in which a particle becomes positively charged by the emission of photoelectrons, occurring when the particles are exposed to solar ultraviolet light. Such a particle can then be driven downward by electrical fields, as for instance by electrical fields generated on a world-wide scale by the interaction of particle streams from the Sun with the Earth's magnetosphere. Local electrical fields extending upwards into the stratosphere may also be generated by intense thunderstorms. Whereas gravity is more effective in causing downward motion the larger a particle (i.e. for bacteria, protozoa etc.) electrical effects are more effective the smaller a particle (i.e. the smaller viruses).

Data on the breakdown of small particles through the stratosphere were obtained from the radioactive isotope rhodium−102 generated in the HARDTACK atmospheric nuclear bomb test of 11 August 1958. The explosion occurred at about 43 km altitude above Johnston Island (16 N, 170 W). The nuclear debris from the explosion went overwhelmingly upwards to heights above 100 km, where the radioactive material spread

7

quickly around the whole world, through largely horizontal motions of the upper air. Because the rhodium–102 was initially in atomic form it took a time of the order of a year to appear at the stratopause, which very likely it did by adhering to small aerosol particles with sizes of the order of the picornaviruses.

After reaching the stratopause the Rh–102 provided a highly effective tracer material for determining the motion of small particles downwards through the stratosphere. Data were obtained from samples of air taken at an altitude of about 20 km, not much above the bottom of the stratosphere. The samples were taken by plane and balloon at various latitudes and longitudes, results being reported by M. I. Kalkstein (*Science*, 137, 1962, 645). Figure A5.1 shows the rising concentrations of Rh–102 that occurred in the northern winter of 1959-60. The primary break down through the stratosphere occurred near latitude 65°N. This was followed from November to March by a major break for the latitudes 45°-50°. (The smaller effect for 15°-30° was probably due to a horizontal spread from higher latitudes.) It was possible to estimate the total amount of Rh–102 produced in the nuclear explosion. Comparing the estimated total with the quantities recovered in the samples it was found that it would take of the order of a decade for the whole of the Rh–102 injected into the stratosphere to reach the tropopause, and thence to ground-level.

Fig. A 5.1.–The fall out of Rh–102 at various latitude intervals from the HARDTACK atmospheric nuclear bomb which was exploded on 11 August 1958.

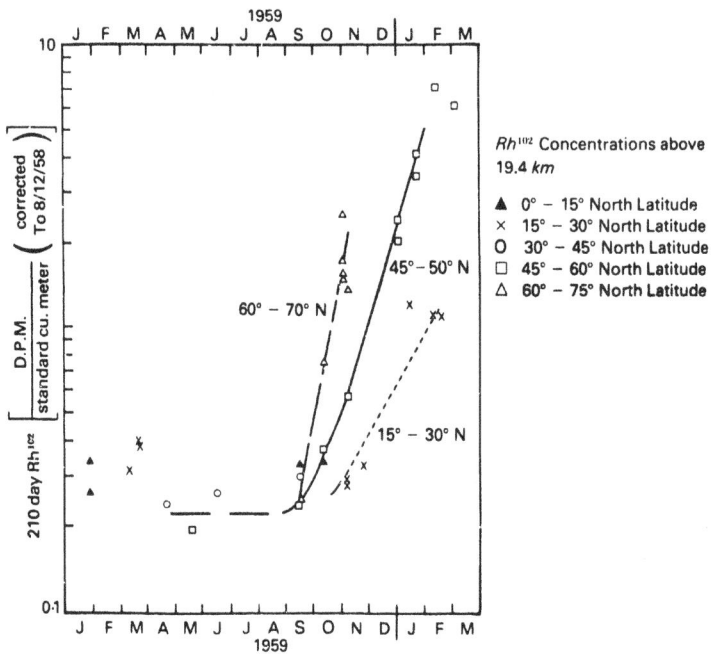

8

Unfortunately, data from the southern hemisphere were much less complete than from the north, but there is little doubt that the middle to late winter rise of Figure A 5.1 would occur also in the south, six months displaced however, from June to September. The winter rise, alternating in the two hemispheres, agrees with the behaviour of a number of viral diseases, of which influenza is a well-known example (entry D 1). In this connection it should be noted that the descent of small particles from near the base of the stratosphere takes place in only two or three weeks, because of strong vertical air movements in the lower troposphere.

The same Winter-Summer effect has been inferred from data of a quite different kind, namely, from measurements of atmospheric ozone concentrations at various heights and various latitudes (c. f. *The Stratosphere* NASA, Dec 1979, page 269). The Earth's atmosphere may be thought of as a heat engine, operating on the difference of temperature between equator and pole. Irrespective of the details of the transfer of the working fluid between the heat source and heat sink, the efficiency of an engine increases with the difference of temperature between source and sink. For the atmosphere, the difference of temperature is greatest in Winter, November to March in the N. Hemisphere and May to September in the S. Hemisphere. Atmospheric *interchanges* between the stratosphere and the troposphere are therefore most vigorous in Winter, with a sixth month alteration between the two geographic hemispheres. The situation in Summer is compartively weak. Indeed in Summer the polar region of the stratosphere actually receives more solar energy than the equatorial region, so that an inversion of the temperature occures. Since the working fluid moves in a heat engine from source to sink, stratospheric air should move towards lower latitudes in Summer and towards higher latitudes in Winter, with the effect (due to the Coriolis force) of generating winds directed east-to-west in Summer and west-to-east in Winter. This accords with the measured zonal winds in the stratosphere.

B

Bacteria and other microorganisms

B 1. The Hardihood of Bacteria

In an article " Extreme Environments : Are There Any Limits to Life " ?, appearing in Comets and the Origin of Life (ed. C. Ponnamperuma) D. Reidel, 1981, D. J. Kushner gives the following table of extreme properties of micro-organisms :

TABLE B 1.1
Some Environmental Limits for Microbial Life

Conditions	Examples of microorganisms that grow in such conditions		
High tempertures (45°C–100°C) (Hot springs, volcanic soils, compost heaps)	Eucaryotic cells Photosynthetic procaryotes Non-photosynthetic procaryotes	less than ,, ,,	60°C 70°C 100°C
Acid hot springs (pH 1, 90°C)	Sulfolubus acidocaldarius		

Conditions	*Examples of microorganisms that grow in such conditions*
Low temperatures (–20°C (?) to 30°C)(Oceans, ice and snow surfaces, Caves, refrigerators and freezers)	Many " psychrophilic " yeast and bacteria
Acid and alkaline conditions (Hot springs, alkaline lakes and soils, some industrial effluents, acid mine waters, coal mine refuse piles)	pH 1–5 Thiobacillus and Acetobacter spp. Eucaryotic algae pH 4–8 Many bacteria and other microorganisms pH 8–11 *Bacillus circulans* and other bacilli *Ectothiorhodospira halophila* ; blue-green algae pH 2–10 *Penecillium* and other fungi
Salt solutions NaCL : 0.01 M 0.3 M 0.3–3.0 M 2.5–5.0 M 0–5. 0 M	Many microorganisms Marine microorganisms Moderate halophiles (some marine) Extreme halophiles Salt tolerant microorgnisms
Low water activity (in concentrated salt or sugar solutions, or on surfaces in dry atmospheres)	a_w : 1.00–0.95 Many microorganisms 0.88–0.75 Extreme halophiles 0.97–0.65 *Xeromyces bisporus* 1.00–0.60 *Saccharomyces rouxii*

Radiation (Dose giving ca. 37% survival for UV (ergs mm^{-2}) and ionizing radiation (kR) in first and second columns respectively

Escherichia call	500	2
Saccharomyces cerevisiae	800	3
Bodo marina	50,000	?
Micrococcus radiodurans	6,000	150

Heavy metals (acid mine waters, laboratory reagents)	Many microorganisms are inhibited by low concentrations (10^{-5} to 10^{-6}M) of heavy metals ; others (e.g. thiobacilli) grow in 1% copper ; some fungi can grow in saturated $CuSO_4$

Definition of water activity

$$a_w = \frac{\text{Vapour pressure of solution}}{\text{Vapour pressure of water}}$$

Kushner argues in the article cited above that the one sure necessity for life is the presence of liquid water, which however need not exist in an open form as in a stream, a pond, or in the ocean. The ability of some microorganisms to replicate at a water activity of as little as 0.6 is achieved by dissolved salts lowering the vapour pressure of the solution to equal the atmospheric water vapour pressure. Thus some **microorganisms** can replicate in an atmosphere of sufficiently high humidity without requiring the presence of pure liquid water. Nor is it essential to consider humidity values in an open atmosphere since humidity values can be considerably increased locally due to capilliarity in tubelike structures. Some microorganisms are able to replicate extremely rapidly when the conditions are appropriate, in a time scale of the order of minute only. Thus a temporary melting of ice can lead to replication, even if the duration of melting is very short, which is a further considerable weakening of the requirement for liquid water. This circumstance is of relevance to the replication of microorganisms in comets, and to the actual presence of microorganisms in meteorites (entry M1).

Likely enough, it is the steam produced when water boils that is the lethal agent for microorganisms. Thus the 100°C upper limit in the above table may well be an artefact due to the terrestrial atmospheric pressure happening to be such that water boils at 100°C. A higher atmospheric pressure might permit survival to a higher temperature. (A recent article by J. A. Baross, shows that this is so.)

B 2. The Resistance of Microorganisms to Ionizing Radiation

X-rays incident on microorganisms cause damage by generating photoelectrons within them. The photoelectrons then cause breaks in the basic genetic structures of organisms, the DNA or RNA in the cases of some viruses. In effect, the process of damage by X-rays is the same as by high speed particles due to radioactivity. The unit measure of damage in the case of X-rays is the rad (r), or the kilorad (kr.). An equivalent amount of damage caused otherwise than by X-rays has the rep (roentgen equivalent physical) as its unit. Here, however, we shall use rads, kilorads, megarads (Mr), as if all damage were caused by X-rays.

The bacterium *Micrococcus radiodurans* with an entry in Table B1.1 was discovered in experiments related the sterilization of meat extracts (by A. W. Anderson, H. C. Nordon, R. F. Cain, G. Parish and D. Duggan in *Food Technology*, 575, 1956). Although 37% of the bacteria were destroyed by a dose of about 150 kr, as shown in Table B1.1, very much larger doses were required to reduce the percentage of viable bacteria below 1%, namely 2500 kr = 2.5 Mr. Thereafter the fraction of survivors fell rapidly with increasing dose, to only 10^{-6} at 5 Mr.

An even more resistant bacterium *Micrococcus radiophilus* was reported by N. F. Lewis (*J. Gen. Microbiol.*, 66, 1971, 29), while a species of *Pseudomonas* has been found in a research nuclear reactor where the average dose was estimated to have been in excess of 1 Mr (E. B. Fowler, C. W. Christenson, E. T. Jurwey and W. D. Schafer, *Nucleonics*, 18, 1960, 102). Even comparatively susceptible species like *E. coli* can be remarkably difficult to knock out completely by ionizing radiation. Thus although only 37% of the strain *E. coli* B/r survived at a dose of 10kr, 1.8% still remained at a dose of 50 kr (R. A. McGrath, R. W. Williams and D. C. Swartzendruber, *Biophysical J.*, 6, 1966, 115). An extreme example of the difficulty of removing the last traces of a microorganism through large doses of ionizing radiation was shown for seemingly unresistant vegatative bacterial by E. A. Christensen (*Acta path. et microbiol. Scandinav.*, 61, 1964, 483), who found that fractions of order 10^{-6} of strains of *Streptococcus faecium* isolated from dust and soil survived at the very high dose level of 2 Mr.

13

Experiments on *Micrococcus radiodurans* are estimated to have caused of the order of 10,000 breaks in the DNA of these bacteria. Yet by a process of snipping and inverse base-copying the bacteria repaired the whole of this enormous damage (c.f. A. Nasim and A. P. James, in *Microbiol Life in Exteme Environments*, ed. D. J. Kushner, Academic Press, 1978). Whole batteries of enzymes are needed to operate such a complex process. This is a case where enzymic requirements are defined externally to the biological system, not relative to the purely internal requirements of the system, a circumstance of importance in calculating the chance of finding enzymes by a random linking of amino acids into polypeptide chains (entry 02).

The dose-rate of ionizing radiation experienced by microorganisms at ground-level is only about 10^{-1}r per year. Although the dose-rate due to radioactive materials in the soil must have been greater than this in the early history of the Earth, a rate of ~ 10 r per year is an upper limit at all times since the Earth possessed an atmosphere. Thus the ability of microorganisms to withstand very high doses that are sudden—worse than being slowly applied—is inexplicable in terms of terrestrial natural selection.

In contrast, the X-ray dose in interplanetary space at times of considerable solar activity can be as high as 10 r *per second*. Hence the ability of microorganisms to withstand very high doses is an important requirement for survival in space over the days that would be required for passage at times of solar activity from safe shielding in the interior of a cometary body to safe shielding below the atmosphere of the Earth (entry B3).

Viruses, and viroids still more, have the advantage of being smaller targets for damaging radiation than bacteria. Thus about 100 kr is needed to produce a single break in the nucleic acid of the smaller viruses, and in excess of 1 Mr for viroids. In addition to this advantage, viruses can use the enzymic apparatus of host cells to repair themselves, even to the astonishing extent of being able to ' cannibalize ', a process in which several inactivated viral particles combine viable portions of themselves to produce a simple active particle.

B 3.   The Resistance of Microorganisms to Ultraviolet Light

Ultraviolet light, especially at wavelengths around 250 nm, destroys the viability of microorganisms. Instead of base-pairs remaining independent of each other as in viable organisms, adjacent bases (not in the same pair) become bonded together and sometimes bonded to proteins, especially when thymine is involved. This has the effect of interfering with the coding of polypeptides and of preventing replication of nucleic acid. Damage due to ultraviolet light does not usually open up breaks in the nuclei acids, however, as does damage by ionizing radiation. Damage due to ultraviolet is mainly reversible, in the sense that if the original chemical bonding can be restored, with the base-pairs returning to their initial relationships, viability is restored. This tends to happen in the presence of visible light through the aid of cell pigments (entry A3). The enzymes involved in this repair process are different from those involved in repairing damage by ionizing radiation.

Figure B3.1 shows the damaging effect of ultraviolet light on three strains of *E. coli*. It is seen that microorganisms are not only considerably variable in their ultra-violet resistance from species to species (Table B1.1) but different strains of the same species can be considerably variable.

Fig. B 3.1. Surviving percentages of various micro-organisms as a function of ultraviolet exposure.

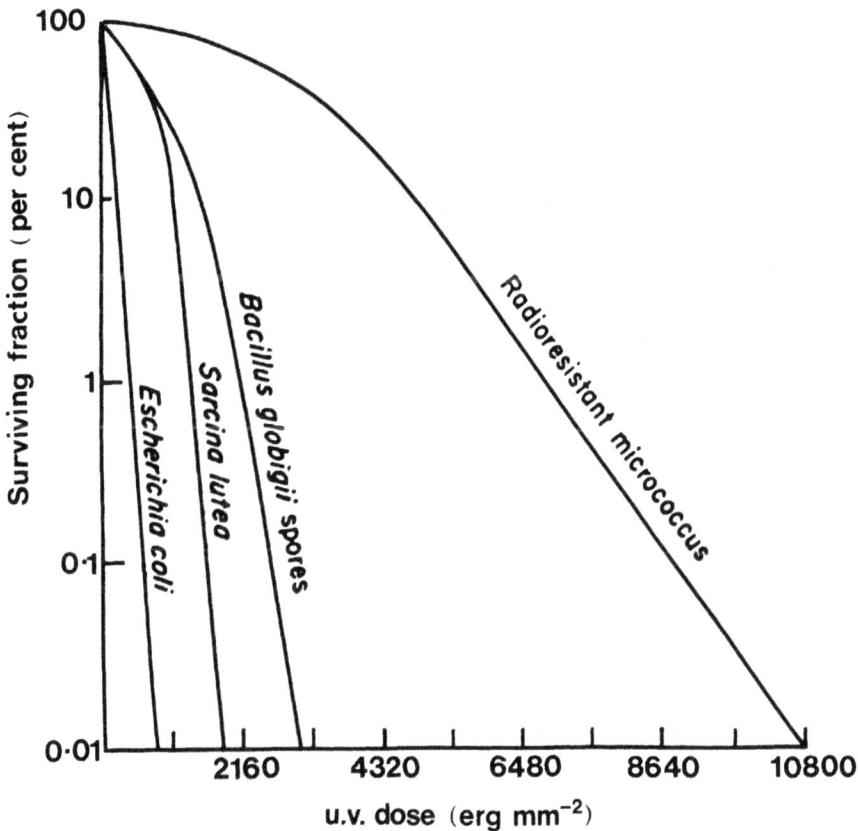

15

Because solar ultraviolet light is essentially completely absorbed by the atmosphere at wavelengths less than 290 nm, terrestrial biology near ground-level is not exposed to the damaging radiation around 250 nm at which laboratory investigations are mostly conducted. There is evidently an evolutionary problem for conventional theory in understanding the great resistance to ultraviolet displayed by some microorganisms, *Bodo marina* for example (Table B1.1). Nasim and James remark (reference in entry B2) :

> " Species vary enormously in their natural resistance to radiation. This variation far exceeds that expected from any differences in the mean levels of radiation to which organisms are exposed. There is, in fact, little or no correlation between natural levels of exposure and species resistance. The reasons for this variation are obscure, but it seems likely that extreme variation is a consequence of some other feature of a species".

The situation is not so mysterious as this passage implies when microorganisms are taken to have arrived at the Earth from space, from evaporated cometary material. The need to survive solar radiation requires resistance to both ionizing radiation and ultraviolet. After arrrival on Earth resistance to high radiation levels becomes inessential, however, and so mutations deleterious to repair processes become tolerated and are permitted to accumulate. Microorganisms that establish long-term residence for themselves here on Earth would therefore be expected to become weaker in their resistance to radiation as time proceeded. The variations referred to by Nasim and James would then be a reflection of residence times : High-resistance organisms would be recent arrivals, low-resistance organisms would be old inhabitants.

One can also argue with plausibility that those minor fractions of the more sensitive organisms (e.g. *E. coli*) which nevertheless survive large dosages achieve their survival because of incomplete spreading of deleterious mutations through the whole bacterial population.

Although an unshielded microorganism outside the Earth's atmosphere would be likely to fall prey eventually to solar radiation, shielding against ultraviolet and soft X-rays could readily be provided. A carbonised layer around a colony of microorganisms with a thickness of 1 μm would be a sufficient shield. This would leave the deleterious effects of only the harder X-rays for which the dose rate, even at times of solar activity, would be less than was discussed in B 2, less than 1 rad per second. Equipped with such a shield, microorganisms could retain viability in space over considerable time intervals, weeks if not indeed months.

B 4. The Disadaptation of Microorganisms

The fraction of microorganisms that interact harmfully with plants and animals is fortunately small. Otherwise we should be overwhelmed by their vast number, $\sim 10^{27}$ bacteria over the whole Earth, with a total mass about comparable with the total mass of the mammals, elephants and all. Ordinary garden soil contains $\sim 10^8$ bacteria per gram, a considerable fraction of farm yard **muck** consists of bacteria, while high bacterial counts have been reported even from deep-sea sediments. On land, microfungi, algae and protozoa are three to four orders of magnitude less numerous than bacteria. In the ocean plankton are dominant, with a total biomass of $\sim 10^{17}$ grams.

It is characteristic of microorganisms that they exist where they can survive, not necessarily where they are well adapted to the local environment. Thermophilic bacteria with optimum temperatures for replication above 45°C exist in arctic soils and waters and in the cold ocean depths (R. H. Mcbee and V. H. Mcbee, *J. of Bacteriology* Vol. 71, 182). **Psychrophylic bacteria with optimum temperatures for replication below 15°C Exist in profusion in warm soils.** A. E. Kriss (in *Marine Microbiology*, **page 91**, Wiley-Interscience, 1962) finds similar bacteria in garden soils and at great pressures in the ocean depths.

These facts fit a picture in which microorganisms with an exceedingly wide range of properties are plastered onto the Earth, some – perhaps the majority – unable to take root, others establishing themselves where they may. The facts do not support the usual evolutionary picture of species becoming well-adapted in competition for available niches. Microorganisms are the raw material of biology, surviving rather than evolving for survival.

It is easy by artificial means to give an opposite impression, by replicating microorganisms in a controlled niche with narrowing parameters. As the constraints are tightened in such a situation species will die by the wayside, until in the limit just one will be left. This last remaining species, or even a particular variety of species, will have the appearance of being closely adapted to the highly restricted environment in question. This evolutionary illusion can even be heightened by exposing the survivors to some mutagenic agent, e.g. to ultraviolet light. The mutagenic agent acts to destroy genetic information, putting various genes out of commission. Paradoxically, this destruction may improve performance in the highly-controlled environment, and the experimenter can all too easily imagine genetic information to have been miraculously created. The illusion here arises for a reason that is well-understood from experience with any complex man-made device. Whenever such a device is required to meet a wide range of specifications, compromises of design become necessary. Should the specifications be made less embracing, some of the previous refinements and precautions become redundant, and if scrapped (genes destroyed) can lead to improved performance with respect to the less embracing specifications.

There is a more valid sense in which bacteria can be said to evolve with respect to their environment, but no creation of genetic information is involved here either. As well as their main circular chromosome, bacteria contain plasmids which carry a few genes. Plasmids behave in several crucial respects like viruses. They emerge from bacterial

17

cells and proceed to 'infect' other bacteria. Since bacteria can generate plasmids from their main chromosome, and since they can also incorporate plasmids into their main chromosome, new genes can be added to a species from other dissimilar species. Severe selection in an artificially controlled environment can through plasmid exchange cause modified species to arise. This is a case of bacterial species taking-in each other's washing, with plasmids forming the raw material rather than the bacteria themselves. Thus plasmids, viruses which operate similarly to plasmids, and the still smaller viroids are more properly to be viewed as the raw material of biology. Perhaps it is even individual genes that should be so regarded.

A well-known example of these apparent selective processes has been the emergence in hospitals of drug-resistant strains of bacteria.

B 5.   The Size-Distribution of Bacteria

The determination of a strictly weighted size-distribution of bacteria would involve a knowledge of the populations of all species. Numbers are so vast, however, that no such reckoning could be made. Criteria of selection are therefore needed. A feasible procedure is to restrict the sample to spore-forming bacteria, for which comparatively good data are available. Even so, the populations of individual species of spore-forming bacteria are not well-determined. However, by counting the number of species of spore-forming bacteria in various size intervals, we can hope to remove variability from species to species through averaging (R. E. Buchanan and N. E. Gibbons, Bergey's Manual of Determinative Bacteriology, The Williams and Wilkins Co., Baltimore). When this is done the histogram shown in Figure B 5.1 is obtained. The abscissa is the diameter in μm for micrococci and is rod-diameter for bacilli. The ordinate is the number of species in the various steps of the histogram. This size distribution will be used later in a discussion of the nature of the interstellar grains (entry 15).

B 6.   Chemoautotrophic Bacteria

Consider the reversible chemical reaction:

$$M_1 + M_2 + \ldots \rightleftarrows N_1 + N_2 + \ldots + Q \qquad\qquad B6.1$$

where M_1, M_2, ... and N_1, N_2 ... are molecules built from a set of atoms A_1, A_2, ..., and Q, taken positive, is the energy released in passing from left to right. The molecules may be present as solids, liquids, or in the gas phase. The thermodynamic balance in such a reaction involves a count of the number of quantum states available to each side of the

18

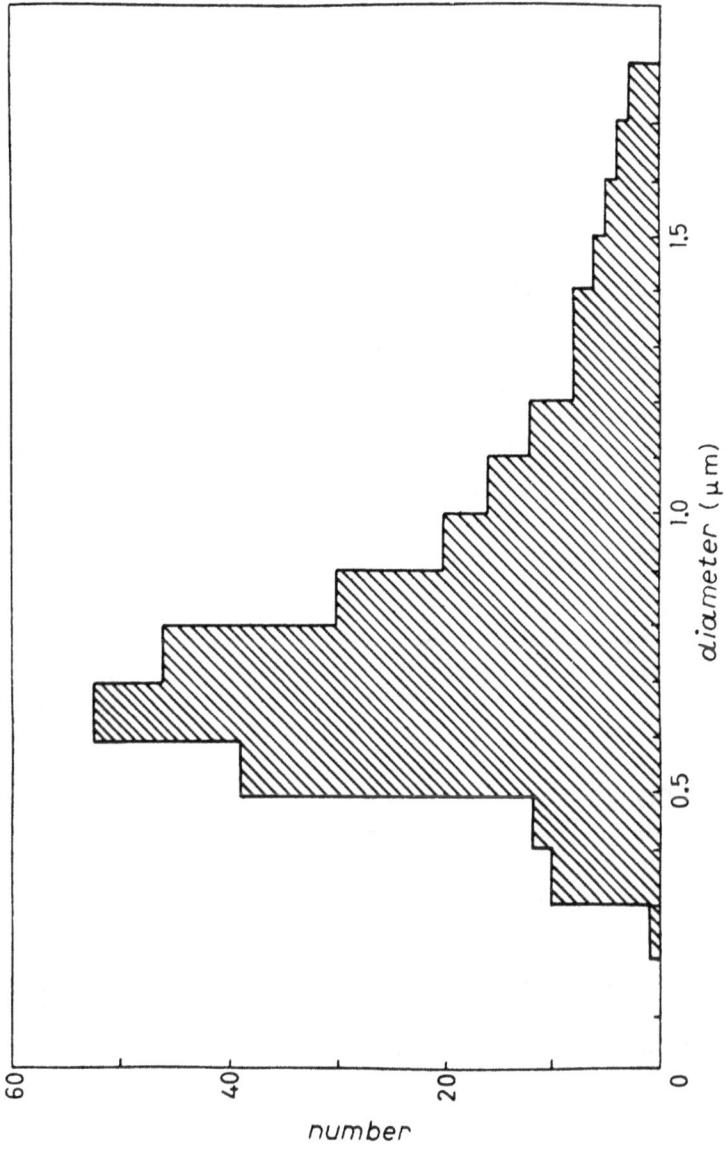

Fig. B 5.1 Size–distribution of spore–forming bacteria.

reaction, as well as depending on the magnitude of Q. Counts of quantum states are usually dominated by the number of gaseous components, with the reaction tending to swing (except at very high pressures) towards the side with most gaseous components. If most gaseous components are on the right in B6.1, then both the energy criterion and the number of states criterion favour a situation in which the atoms A_1, A_2, . . . are combined in the molecules N_1, N_2, . . . If, however, most gaseous components are on the left in B6.1 we have opposing tendencies. The energy criterion ($Q > 0$) favours the right-hand side and the number of states criterion favours the left-hand side. The latter is not appreciably temperature-dependent, whereas the energy criterion is sensitively temperature-dependent through the well-known Boltzmann factor exp $(-Q/kT)$. At high enough temperatures this factor is weak and the reaction therefore swings to the left, toward the set of molecules with most gaseous components. As the temperature declines the Boltzmann factor becomes more dominant, however, until for low enough temperatures it dominates the thermodynamic situation, and the reaction then swings left-to-right, provided the temperature is not so much reduced that the kinetic rate of the reaction has become too slow for thermodynamic equilibrium to be attained, in which case the atoms A_1, A_2, . . . stay in the molecular configuration M_1, M_2, . . . **out of thermodynamic balance.**

As an example, consider the reaction :

$$CO_2 + 4 H_2 \rightleftarrows CH_4 + 2 H_2O + Q$$

B6.2

with all components gaseous, where $Q \cong 40 \text{ k cal}$ per mole of CH_4 produced. At low enough temperatures, for instance at room temperature, the value of Q favours the right hand side of B6.2. Yet if one takes $CO_2 + 4 H_2$ in the laboratory under strictly inorganic conditions, the mixture can be kept for an eternity without the reaction swinging left-to-right. The problem lies with the strong binding of oxygen to carbon, especially the binding of CO. To offset the large amount of energy required to pry the oxygen atom loose from the carbon, it is necessary to bring the four atoms of hydrogen in CH_4 simultaneously together with the carbon. This condition under inorganic conditions is highly improbable, making the conversion to CH_4 exceedingly slow.

It is in such conditions that the chemoautotrophs operate. The methanogens are bacteria that catalyse the reaction B6.2 from left-to-right. Catalysis works through a series of intermediate steps in which the energy exchanges are much smaller than if one were to attempt B6.2 all in one step. The situation is like pumping water uphill in many small steps, eventually establishing a kind of siphon situation with the process flowing smoothly from left-to-right.

The chemoautotrophs are an exceedingly numerous class of bacterium, so numerous that it is tempting to generalize by saying that, wherever a chemical reaction of the kind described above is to be found, there will be a chemoautotroph living on it.

Their profit is the energy Q. The chemoautotrophs may very properly be described as the scavengers of thermodynamics, the raison d'être for their existence being the extreme slowness with which thermodynamic equilibrium is achieved non-biologically in a variety of energy-producing chemical reactions. *If the reactions in question could proceed at appreciable speed non-biologically there would be no niche for the chemoautotrophs.*

Inverting this last remark, wherever we find the products of a chemoautotroph, as for instance the product CH_4 of the methanogens, it is plausible to suppose that a bacterium has been at work in order to generate the observed product. In terrestrial situations such a prediction can often be put to an explicit test, and so far as we are aware the outcome is always positive. In non-terrestrial applications of this deduction an explicit test may not be possible, and then the inferred presence of bacteria is of great interest.

21

C

Comets

C 1. The Composition of the Volatile Fraction of Comets

Table C1 . 1 shows the ratios of numbers of hydrogen, carbon, nitrogen and oxygen atoms – the four main life-forming elements – present in material samples of various kinds.

TABLE C 1.1

Ratios of Numbers of Atoms

Element Ratio	Cosmic Material	Earth's Biosphere*	Bacteria**	Mammals**	Volatile Fraction of Comets**
H/O	1100	1.7	2.2	2.3	1.8
C/O	0.447	0.0056	0.22	0.40	0.32
N/O	0.126	0.0035	0.05	0.09	0.08

*The Earth's biosphere is defined here to be the atmosphere, the ocean, and a 1 kilometre surface layer of continental rock. This gives the nearest, although still a poor, approximation to the H,C,N,O ratios in living material. The atmosphere and oceans alone give a still worse C/O ratio, while including more rock gives worse H/O and N/O values.

**After A. H. Delsemme, determined spectroscopically for material evaporated from comets.

A high fraction of cometary material is volatile. Thus comets are basically of the same H,C,N,O composition as living forms, a circumstance explained most simply by the existence of high concentrations of microorganisms in comets. Cosmic material, as it exists in the Sun for example, is not at all lifelike in its composition. Nor are any other bodies except comets, an important point first emphasised by A. H. Delsemme.

ᗨ 2.   The Size Distribution of Cometary Material and its Relation to the Arrival of Microorganisms on the Earth

It would be desirable to have detailed knowledge of the amount of material that enters the Earth's atmosphere, particularly with respect to the size distribution of the material ; that is to say the average amount per year of chunks of material arriving with sizes in the range from a_1 to a_2 for any specified values of a_1 and a_2. Unfortunately we have no such precise information, but the following simple formula appears to fit the known facts as an acceptable rough approximation. The rate of arrival is

$$\sim 100 \; \ell n \; a_2/a_1 \; \text{tons per year} \qquad\qquad C2.1$$

Full-scale comets with masses $\sim 10^{10}$ tons, say from 3.10^9 to 3.10^{10} tons corresponding to $\ell n \; a_2/a_1 \cong 1$, have been estimated to hit the Earth about once in 100 million years, giving an averaged annual rate of ~ 100 tons, in agreement with C2.1. Micrometeorites ranging in size from ~ 10 nm up to 0.1 mm correspond to $\ell n \; a_2/a_1 \cong 10$, so that C 2.1 gives $\sim$ 2000 tons per year for such a range of micrometeorites, in agreement with the order of magnitude discussed in entry A 2.

Formula C 2.1 predicts that about one body in the mass range from 30 to 300 tons should enter the atmosphere each year. Such a body would cause an extended shower of meteorites. With most such showers falling into the ocean and in remote areas of the land we might expect actually to observe about one shower per decade. Several at the lower end of this scale have been observed over the past century.

Formula C 2.1 predicts the arrival of one or two bodies per century with masses from 1000 to 10,000 tons. Very likely it was such a body that caused the Tunguska disaster in Siberia in 1908.

Considering bodies in a unit size range a_1 to a_2 (defined by $\ell n \; a_2/a_1 = 1$) those with masses around M tons enter the atmosphere every ~M/100 years. The number of such bodies, not entering the atmosphere but passing us by within n Earth radii, would be :

$$\sim 100 \; n^2/M \; \text{per year} \qquad\qquad C 2.2$$

If we set the distance of passing at half the radius of Earth's orbit around the Sun, then $n \simeq 10^4$, and C 2.2 gives one body per year for $M = 10^{10}$ tons, just the full blown cometary case. However, C 2.2 also gives many closer passages for bodies of smaller M. Thus with $n \cong 10^3$ we have 100 passages per year for bodies with masses of $\sim 10^6$ tons, and for $n \cong 10^2$ we have 100 passages per year of bodies with masses $\sim 10^4$ tons. According to C 2.2 there are sizeable chunks of material passing frequently at comparatively close distances to the Earth, bodies that would produce devastation similar to the Siberian meteorite of 1908 if they chanced to hit the Earth.

Bodies with masses in the range from $\sim 10^4$ to $\sim 10^6$ tons could arise from the break-up of larger full-sized comets. They could be debris from collisions of comets with asteroids, in which case they could be bits from either comets or asteroids, or they could be primary cometary material coming directly inwards from the far-distant outer regions of the solar system. There seems no reason why at least a fraction of the smaller bodies should not be of primary origin, when they will have a composition like full-sized comets (entry C 1) and will experience evaporation due to solar heating, certainly on a smaller scale but in a similar fashion to the evaporation of the comas and tails of visible comets.

The evaporated cloud from any cometary body presents a far greater target area to the Earth's atmosphere than does the compact body itself, making it inevitable for the Earth to encounter **evaporated** material from many of the smaller bodies which pass close by. Indeed the Earth may be thought of as being permanently immersed in a diffuse halo of cometary material, some of it from distant full-blown comets but a fair fraction of it from those bodies with masses in the range $\sim 10^4$ to $\sim 10^6$ tons which pass by not far from the Earth. This picture agrees with an investigation by Z. **Sekanina** (*Icarus*, 13, 1970, 475) who analysed 19,303 meteor trails having paths explicitly determined by the 6-station Radio Meteor Project at Havana, Illinois. It was found that while some of the meteor paths could be associated with the orbits of known full-blown comets, the majority of the paths could not be so related. This was consistent with a considerable fraction of the meteors being debris from smaller bodies passing close by the Earth.

Microorganisms evaporated from a body passing by at a distance of 100 Earth radii, 6.4×10^{10} cm, and travelling at a speed of ~ 40 km per second relative to the Earth, could reach the Earth in only 1.6×10^4 seconds = 4.4 hours. Thus the exposure of such a micro-organism to solar X-rays and solar ultraviolet light (entries B2 and B3) could be quite brief. Microorganisms from a full-scale comet at a distance of 10,000 Earth radii would, on the other hand, be required to withstand solar radiation for a hundred times as long, about 20 days. It is evidently more favourable, especially for any radiation-sensitive microorganisms, to pass from the shelter of a smaller close-by cometary body to the shelter of the Earth's atmosphere than it would be to arrive from a far more distant full-blown comet. The present considerations show there is no requirement for microorganisms reaching the Earth to have had a long residence in interplanetary space.

C 3. Observational Data for Particles Evaporated from Comets

An investigation by K. Saito, S. Isobe, K. Nishioka and T. Ishii, reported in *Icarus*, 47, 1981, 351, finds a minimum value of 0.4 for the ratio of solar gravity to radiation pressure experienced by particles evaporated from Comet West and Comet Mřkos. A similar result has also been recently quoted from **Halley's comet in 1910** (Sekanina, 1981). According to entry C6, this ratio implies the presence of graphitised bacteria in the comet tails.

25

Figure C3.1 shows broadband infrared photometry of radiation from particulate material evaporated by the **comets** Tuttle, Meier, Stephan-Oterma and Bowell, together with solar emission in the J, H, K, and L bands. (The measured values have been normalized with respect to the J band). The heliocentric distances in astronomical units of the comets at the time the observations were made are given in the figure legend. The authors of this work, M. G. A'Hearn, E. Dwek and A. T. Tokunaga (*Astrophys. J.*, 248, 1981, L147) also obtained narrower-band photometry of Tuttle and Stephan-Oterma, from which they concluded that the particulate material cannot consist of common ices (H_2O, CO_2, CH_4 or NH_3). The same conclusion follows already from Figure C3.1. The cometary measurements in the J, H and K bands are relatively the same in spite of the considerable variability of their heliocentric distances, r = 1.16 for Tuttle and r = 5.62 for Bowell. This shows two things, that the radiation in the J, H and K bands comes from

Fig. C 3.1.   *Relative flux densities of several comets at JHKL, at various perihelion distances.*

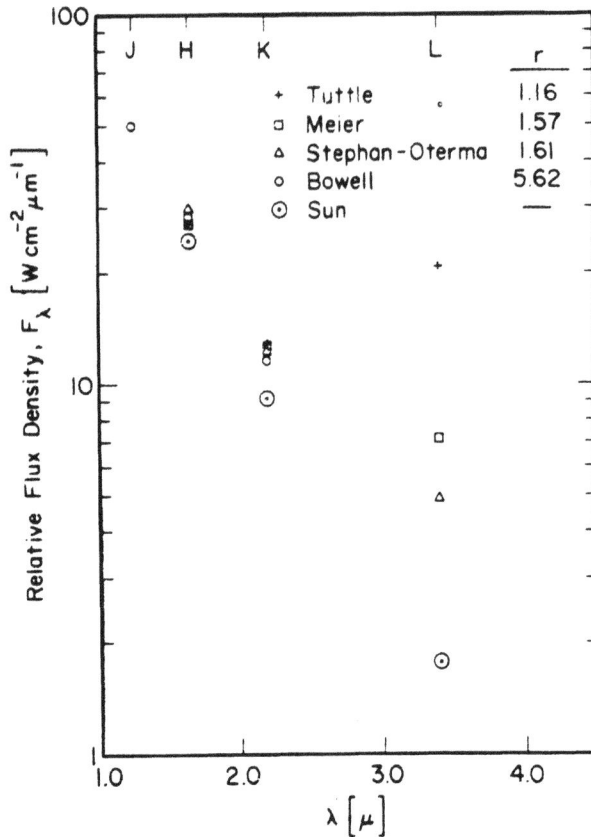

a reflection of solar radiation, and that the reflective properties are the same for the particles evaporated from all four comets. The latter could not be true for common ices which might persist at r = 5.62 (H_2O would persist) but which would surely be evaporated at r = 1.16.

The radiation measured in the L band is strongly dependent on r, however, showing the L-band radiation arises mainly from intrinsic emission by the cometary particles rather than from a reflection of solar radiation. The cometary particles absorb some fraction of the visible sunlight incident upon them, which raises their temperatures until an energy balance is attained with emission in the infrared equalling the energy of the absorbed visible radiation. Small micron-sized particles in such a situation emit their infrared radiation mainly at those wavelengths where they have the largest oscillator strengths, not crudely as black bodies. Thus the emission in the L band, quite intense for Tuttle with the least heliocentric distance, shows appreciable oscillator strengths at wavelengths within the L band, i.e. at wavelengths around 3.4 μm. Such oscillator strengths are characteristic of the C-H stretching mode found in all organic materials. While these measurements do not have the resolution to establish the presence of explicitly biological material (entry 16) they suggest that cometary particles are composed mainly of organic material.

C 4. Cometary Periodicities of Relevance to the Incidence of Microorganisms onto the Earth

The cometary material that we see today has been stored over most of the history of the Earth in the far-distant regions of the solar system. Over most of the past 4.6×10^9 years such material, whether in full-blown comets or in the smaller bodies discussed in entry C2, has followed orbits with aphelion distances of a few tenths of a light year and with perihelion distances that were probably comparable to the radii of the orbits of the planets Uranus and Neptune. Bodies in such orbits have a probability at each perihelion passage of a few parts in a million of experiencing a sufficiently close encounter with Uranus or Neptune to change their orbits appreciably. The change may add appreciable energy to a body, in which case its orbit becomes hyperbolic and it leaves the solar system entirely. Or the body may lose appreciable energy, when the aphelion distance of its orbit is reduced, with a consequent shortening of its period of revolution about the Sun, a shortening of period that brings the comet more frequently into the region of the planets Uranus and Neptune, thereby increasing the number of its perihelion passages, increasing the chance of the process being repeated, with a further planetary encounter either ejecting the body entirely from the solar system or shortening its period of revolution about the Sun still further.

27

If it were not for the presence of the still more massive planets, Saturn and Jupiter, lying inside the orbits of Neptune and Uranus, the evolution of the cometary orbit described in the previous paragraph would cease with the cometary body either ejected entirely from the solar system or with its aphelion distance reduced so as to be within the orbit of Uranus. But Saturn and Jupiter act in the latter case in the same way that Uranus and Neptune did to begin with, either causing the cometary body to be lost from the solar system or reducing its aphelion distance and period of revolution still further, until ultimately the aphelion distance comes inside the orbit of Jupiter and the period of revolution is reduced to only a few years. This has become the situation with several of the so-called short-period comets. Others of the short period comets are in the last stages of this process.

Comets lose material more and more rapidly by evaporation as their periods of revolution decrease, because they come to perihelion more often and so experience heating by the Sun with increasing frequency. It follows that if we are concerned with the evaporation of microorganisms from cometary bodies the short-period evolved orbits will be of particular relevance. Long-period full-blown comets which happen to have perihelion distances less than 1 astronomical unit will on exceptional occasions have effects that are very great, but if we are concerned with the everyday situation it is to the short-period smaller bodies that we should look, especially to those smaller bodies with orbits that nearly intersect the Earth's orbit, since it is these that can on occasion pass by the Earth at comparatively small distances, permitting easy transit of microorganisms from the cometary bodies to the atmosphere of the Earth (entry C2).

To fix the situation more closely, let us consider a cometary body with a period of three to four years, and with an orbit that nearly intersects the Earth's orbit, a body which on occasions passes close by the Earth. Individual microorganisms with sizes ~ 1 μm or less evaporated from the body are quickly whipped away by solar radiation pressure, and it will only be on the occasions when the Earth and the body are close to each other that capture of organisms by the Earth's atmosphere is possible. Since in general the orbital periods of the body and the Earth will be incommensurate, such occasions for a particular body must be rare, perhaps separated by many centuries, much longer than the orbital period of the body itself. For evaporated colonies of microorganisms with dimensions of 10 μm or more radiation pressure is, on the other hand, considerably smaller than solar gravity. Even so, radiation pressure prevents the orbit of a colony from being closely the same as that of the body from which it was derived. The effect is to produce variability in the orbits of individual colonies, causing a diffuse halo of colonies to accompany the body in its orbit. The halo would spread more or less indefinitely for inorganic particles, but if we confine attention to colonies of microorganisms that retain viability, the extent of the halo will be determined by the length of time the microorganisms can resist the deleterious effects of the harder solar X-rays (the interiors of colonies would be satisfactorily protected against soft X-rays and ultraviolet light—entries B2 and B3). The more sensitive the colonies, the smaller the viable halo accompanying the cometary body and the less frequently will the Earth pass through it. The less sensitive the microorganisms, the more the viable halo will extend itself along the orbit of the body and the more frequently the Earth will pass through it. Indeed, for a sufficient extension of the halo the Earth encounters it with an average period equal to the period of the cometary body itself. These statements will become clearer from a consideration of a few examples which show the curious effects that can arise from the mathematical effects of incommensurability.

Consider the idealised problem illustrated in Figure C4.1. Think of the heavily marked sector of the circle (subtending angle θ at the centre) as a band of particles which moves uniformly around the circle in P years, where P is taken to an irrational number. The point A is fixed on the circle, the situation shown in Figure C4.1 being the initial moment, t = 0. An observer visits point A for a brief moment at precisely yearly intervals, starting with a visit at t = 0. In which years does the observer find himself within the band of particles ?

Fig. C 4.1.

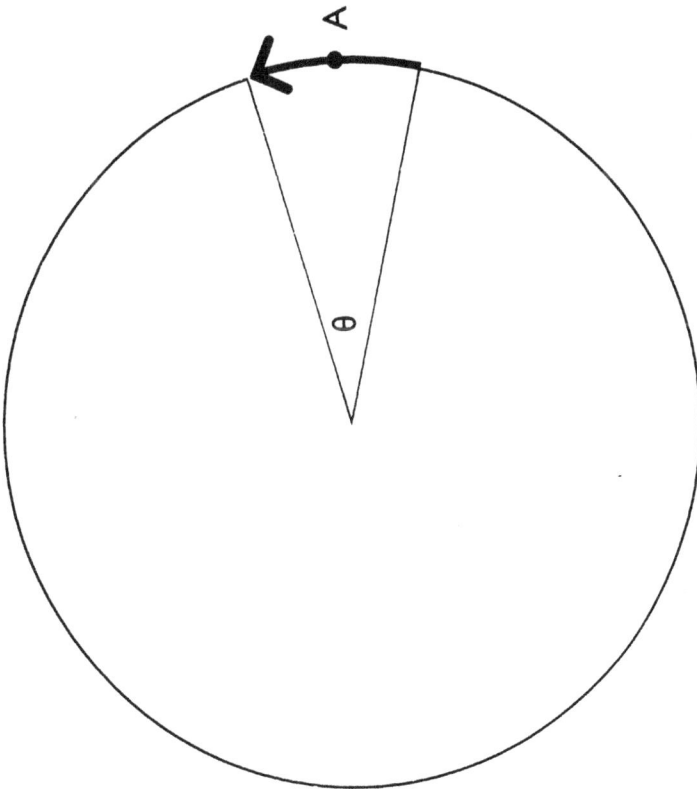

29

If we think of the circle of Figure C4.1 as the auxilliary circle of the elliptic orbit followed by a cometary body, and of the visits of the observer to point A as the intersection of the Earth's orbit with the cometary orbit, the analogy to the physical situation will be clear. For θ small, the mathematical problem is very similar to the physical problem, while even for θ comparatively large the essential features of the incommensurability implied by the choice of an irrational value for P are the same.

To obtain a cometary periodicity between three and four years start with an appropriate ratio of two prime numbers, say $797/223 = 3.57\ldots$ Although it takes a while before the digits repeat in the latter decimal representation, since $797/223$ is a rational number there is still **commensurability** with the 1 year periodicity of the earth—indeed the situation repeats itself exactly in 797 years. To ensure incommensurability add to $797/223$ a small integral fraction of π, for example

$$P = (\frac{797}{223} + \frac{\pi}{300}) \text{ years} \qquad\qquad \text{C 4.1}$$

It is then found that for $\theta = 2°$ the observer finds himself within the belt of particles in the following years :

$$t = 0, 233, 276, 509, 552, 785, 1061, 1294, 1337, \ldots$$

Now notice the dramatic effect of increasing the extent of the belt of particles. For $\theta = 90°$ the observer is within the belt in the following much more closely-spaced years :

$$t = 0, 4, 7, 11, 14, 18, 25, 29, 32, 36, 39, 43, \ldots$$

The sequence for each choice of θ contains apparent regularities. Jumps of 233 years keep appearing in $t = 0, 233, 276, 509, \ldots$, and given such a sequence one might have the impression that 233 years was an important feature of the physical phenomenon involved, whereas the 233 year jumps are almost incidental, an artefact of the precise choice of θ. It is clear that the incommensurability of P requires the average separation in the above sequences to be approximately $2\pi/\theta$, with the approximation improving as the time spans of the sequences increase. However, the average $2\pi/\theta$ is not achieved in any obviously simple way.

When the halo structure is compact, the false apparent periodicities are sensitively dependent also on the **precise** choice of P. Thus again for $\theta = 2°$, but with P slightly changed to

$$P = (\frac{797}{223} + \frac{\pi^3}{300}) \text{ years}, \qquad\qquad \text{C 4.2}$$

the observer finds himself inside the halo of particles in the following years :

t=0, 114, 228, 342, 456, 1151, 1265, 1379,. . . .

the regular jumps now being 114 years, instead of the jumps of 233 years when P is given by C4.1.

For $\theta = 20°$ and P given by C 4.2, the observer is within the halo of particles in the years.

t = 0, 11, 22, 33, 81, 92,. . . .

when a false apparent periodicity of 11 years has appeared. This simple mathematical example gives a savage warning against attempting to infer the nature of a phenomenon from an observed apparent periodicity, since the observed regularity may be only a product of very minor details in the physical situation itself.

C 5. The Rate of Evaporation of Water-based Volatile Materials from Cometary Bodies

Of the ' ices ' which are usually considered as possible components of cometary material, CH_4, NH_3 CO_2 and H_2O, we are concerned here only with water-based material. Methane, ammonia and carbon dioxide are all more volatile than water, and if there was any of them present initially we consider the CH_4, NH_4 and CO_2 to have been evaporated already from the body.

Talking account of orientation effects, and of an element of the surface of a rotating body being as much turned away from the Sun as towards it, the time average of sunlight incident on such an element is about 300 calories per $metre^2$ per second at a perihelion passage at a distance, say, of half an astronomical unit from the Sun. The total energy received for the whole perihelion passage, occupying about a month, is thus $\sim 10^9$ calories per m^2 of the surface of a cometary body. Most of the energy of sunlight is in the visible region of the spectrum, and visible light penetrates about 10 metres into smooth ice. Hence the energy of the visible component of sunlight, $\sim 10^9$ calories per m^2 of the surface of the body, is distributed throughout a layer about 10 metres deep, for an average energy absorption in the layer of $\sim 10^8$ calories per m^3, or ~ 100 calories per cm^3, about sufficient to melt the ice but not to evaporate it.

The absorbed energy is not cumulative from one perihelion passage to the next, because the energy is radiated over the several years which the body spends near aphelion. Thus an icy surface layer about 10 metres thick becomes hard frozen near

31

aphelion without the internal small bubbles sometimes found in terrestrial ice and tends to melt near perihelion, with little evaporation occurring due to the absorption of visible light, the main energy component of sunlight.

The situation is quite different for the infrared component of sunlight. Starting at a wavelength of about 1 μm, water has strong absorption bands going into the infrared. Solar radiation in these bands is absorbed in a quite thin layer at the surface of a cometary body, and this energy (because of the thinness of the layer) is available to supply the latent heat of evaporation of water. Taking the energy in the water absorption bands to be 25% of the total solar radiation, the amount available for evaporation is ~ 2.10^8 calories per m^2 of surface for each perihelion passage of a cometary body, sufficient to evaporate a depth of about 30 cm of water.

From the above estimates it follows that a body with an initial radius of 30 metres, typical of the smaller cometary bodies considered in entry C4, would suffer evaporation of its icy materials in about a hundred perihelion passages. With a period of 10 years for the orbital revolution about the Sun, typical for a short-period cometary body, the watery material would therefore be lost in about a millennium. Remaining behind would be the non-volatile materials of the body, likely enough with a composition similar to the carbonaceous chondritic meteorites. Recent data on the nature of the carbonaceous material in a particular member of this class of meteorite will be given in entry M1.

C 6.   Radiation Pressure on Microorganisms of Cometary Origin

The ratio R of the solar gravitational force to the pressure of solar radiation on small particles is in general a complicated function of particle size and of the optical constants of the grain material. It is calculated from the expression.

$$\frac{R}{\rho} = \frac{1.71 \ a}{<Q_{pr}>} \ , \qquad\qquad (C6.1)$$

where $<Q_{pr}> = <Q_{ext} - \overline{\cos\theta} \ Q_{sca}>$ is the efficiency factor for radiation pressure given by the Mie formulae: averaged over the solar spectrum, a is the radius of spherical particles, and ρ is the density of the grain material.

Fig. C6.1 shows this ratio for hollow micro-organisms with a refractive index value n = 1.167 (for a discussion of this value see entry I 5) and for various values of the absorptive index k. We note that in the case of purely dielectric non-absorbing bacteria solar gravity must always exceed the radiation pressure by a significant factor. The situation is dramatically altered for non-zero values of k, however. With an absorptive index k = 0.05 the minimum value of R becomes less than unity for $\rho \simeq 0.6$ g cm^{-3}, a density value appropriate for vegetative bacteria from which water has been evaporated.

The absorptive indices of bacteria due to the presence of pigments such as carotenoids remains in general smaller than 0.01. At such values of k the visual mass absorption coefficient of grains at optical wavelengths remains less than ~ 1000 cm^2 g^{-1}, the value appropriate for infrared wavelengths at ~ 6 μm and 10 μm. Absorptive indices $k \gtrsim 0.03$ implies a mass absorption coefficient $\kappa \sim 4.3 \times 10^3$ cm^2 g^{-1} for bacterial grains (n=1.167), and this is achieved by graphitisation to only about 1/20 of the grain radius.

Carbonisation, when it occurs, would lead to a run-away situation, with biomaterial going all the way to graphite unless the particles become expelled quickly enough from the vicinity of the Sun. This is because the absorption coefficient at visual wavelengths overshoots the value at infrared wavelengths, leading to a strong green-house effect within the grain. For absorptive particles small compared to the electromagnetic wavelength the cross-section for radiation pressure is very nearly that for true absorption. In this case the ratio of gravity to radiation pressure is

$$\alpha \simeq \frac{4\pi \, G \, Mc}{\kappa L} \qquad\qquad (C6.2)$$

where κ is the mean mass absorption coefficient of the grains and M, L are the mass and luminosity respectively of the Sun. Observational data for comets Mřkos and Halley amongest others lead to a value $\alpha = 0.4$ for the most highly accelerated grains. The value $\alpha = 0.4$ leads from (C6.2) to $\kappa \simeq 30,000$ cm^2 g^{-1} , which is almost precisely the visual mass absorption coefficient of graphite particles of sufficiently small radius, or of hollow particles with sufficiently thin skins. The visual mass absorption coefficient of iron, on the other hand, is about six times lower giving maximum acceleration values that are six times lower than the observations show for comets' tails, so that the accelerated grains cannot be iron, or any other metal.

Fig. C 6.1 The ratio of solar gravity to radiation pressure per unit bulk density (in g cm^{-3}) for spherical grains with n = 1.167 and for varous values of k.

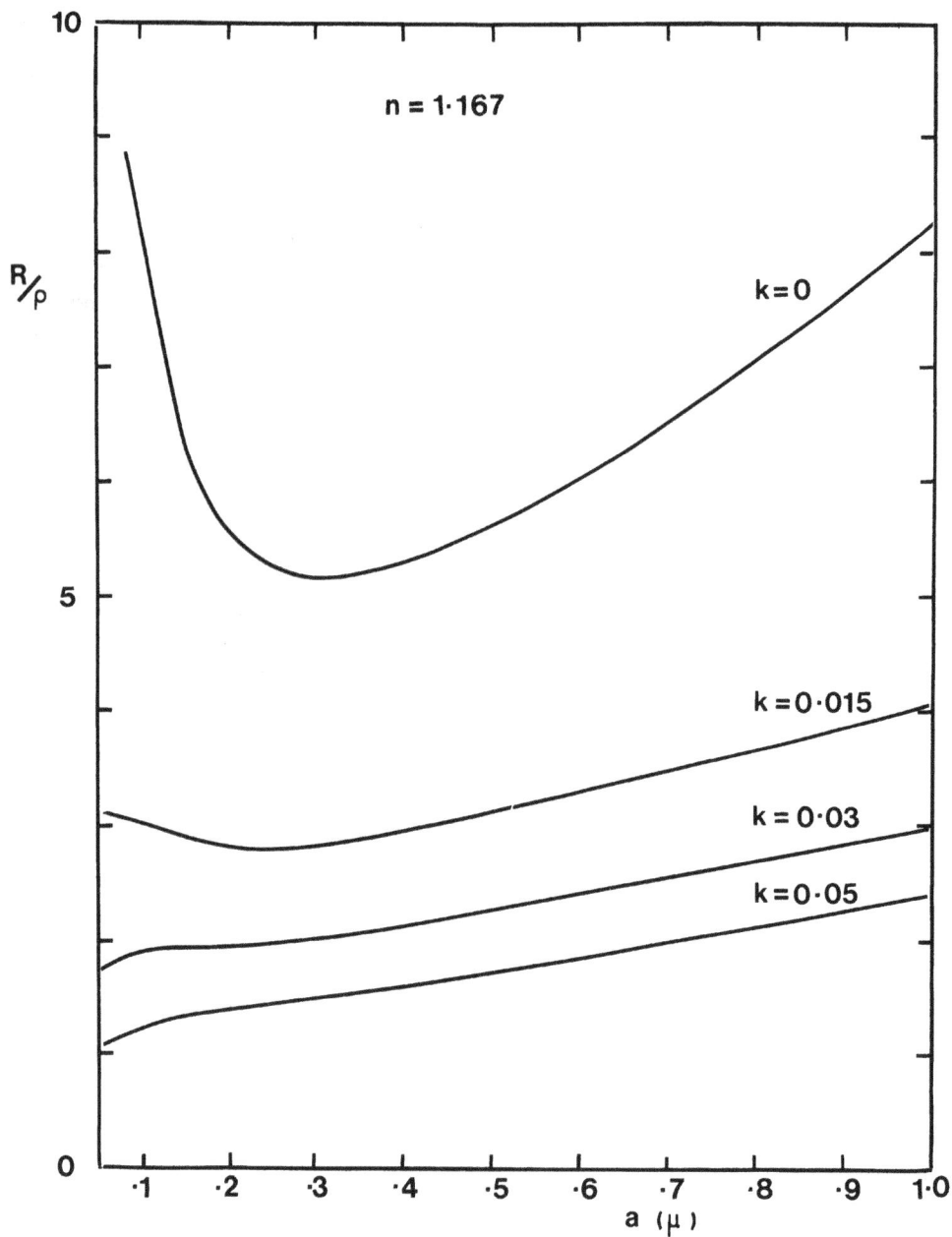

It would seem a property unique to micro-organisms that they lead when sufficiently heated in vacuum to carbonisation and that in a carbonised condition they have a mass absorption coefficient that is exactly right to account for the minimum observed values of α in comets, and that this minimum is repeated almost precisely from one comet to another. For inorganic grains no such natural explanation exists, although of course it **might** be said that the n and k values of a particular subset of the grain population happened by accident to imitate the properties of graphite.

Figure C6.2 shows the ratio of solar gravity to radiation pressure (per unit grain density) for silicate grains. With a silicate density of $\rho \sim 3$ g cm^{-3} we see that there is no reasonable case for which R could be less than unity. Silicate grains cannot threfore explain the repulsion observed in comet dust tails.

Fig. C 6.2 The ratio of solar gravity to radiation pressure per unit bulk density (in g cm^{-1}) for spherical grains with n = 1.65 and for vaious values of k.

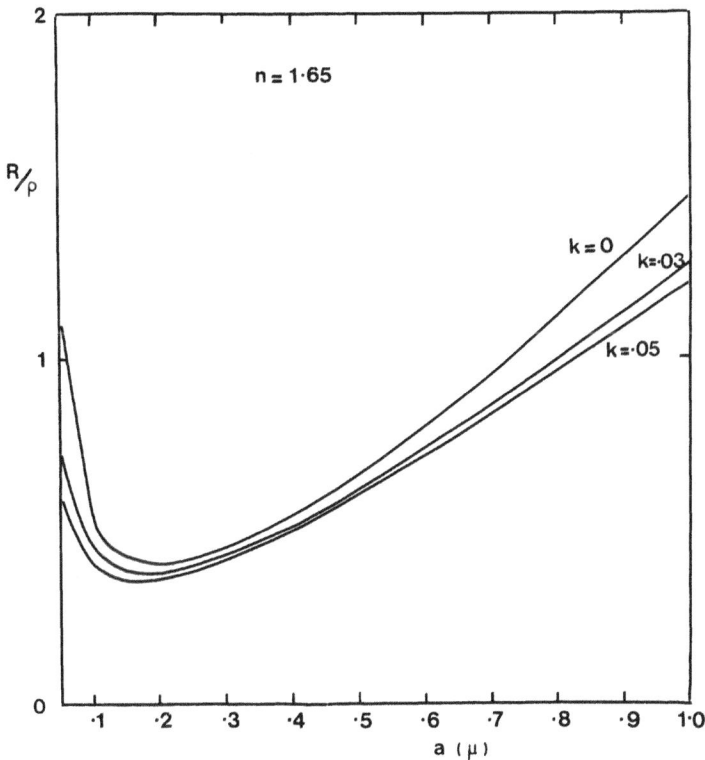

D

Diseases

D 1. The Arrival of Space-Borne Microorganisms at the Earth's Surface

Do we have proof that microorganisms are being added to the Earth from outside ? The problem in seeking an answer to this question is to distinguish new microorganisms coming from outside from the large populations that are in residence here already. The best chance of coping with this difficulty would be if some among the new microorganisms were able to cause diseases in terrestrial plants and animals. This is because a pathogen multiplies itself enormously in the body of its victim, in some cases by thousands of billions. Terrestrial plants and animals can therefore be viewed as highly sensitive detectors for pathogens from space, although of course there is still the problem of distinguishing attacking microorganisms of external origin from attacks due to pathogens already in residence here.

There are many situations, both historically (entries D5, D7 and D8) and in modern times, that are suggestive of the arrival of space-borne pathogens. Here is a quotation from Sir Christopher Andrewes' book *The Common Cold* (Weidenfeld and Nicolson, 1965) :

> " Van Loghem in 1925-26 carried out a postal canvas of about 7,000 persons in different parts of the Netherlands over a period from September to June. He was concerned to find out about the occurrence of colds in relation to time and space. He analysed the results of his canvas and plotted them

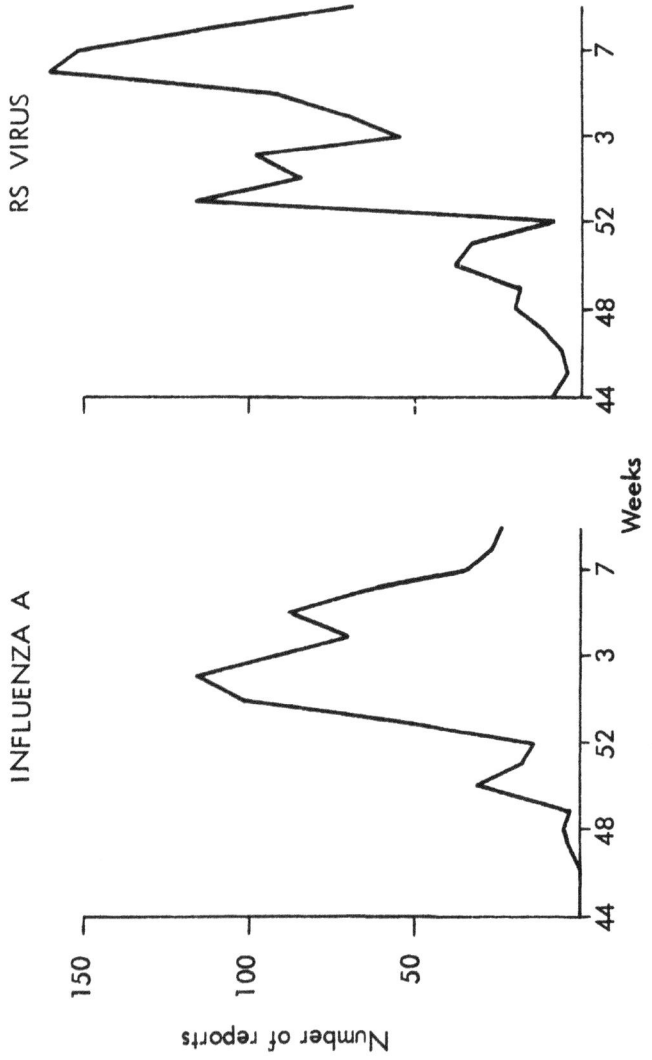

Fig. D 1.1. Comparison of influenza A and RS virus over the winter of 1980/81 (CDR81/09).

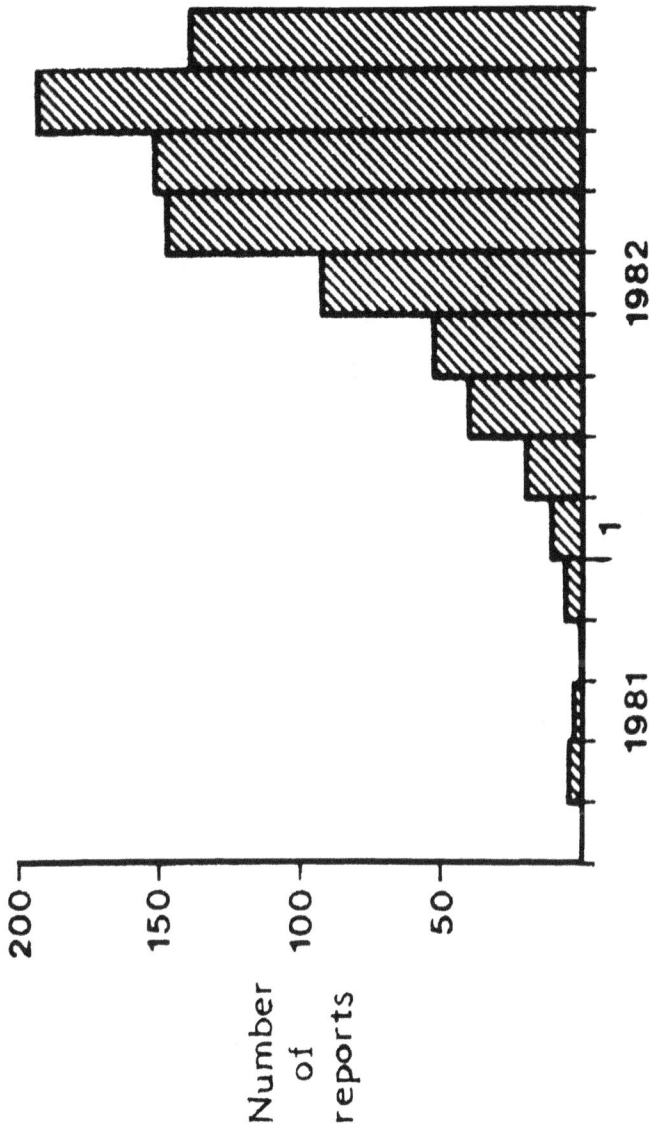

Fig. D 1.2. Influenza B for the winter of 1981/82 (CDR 82/09).

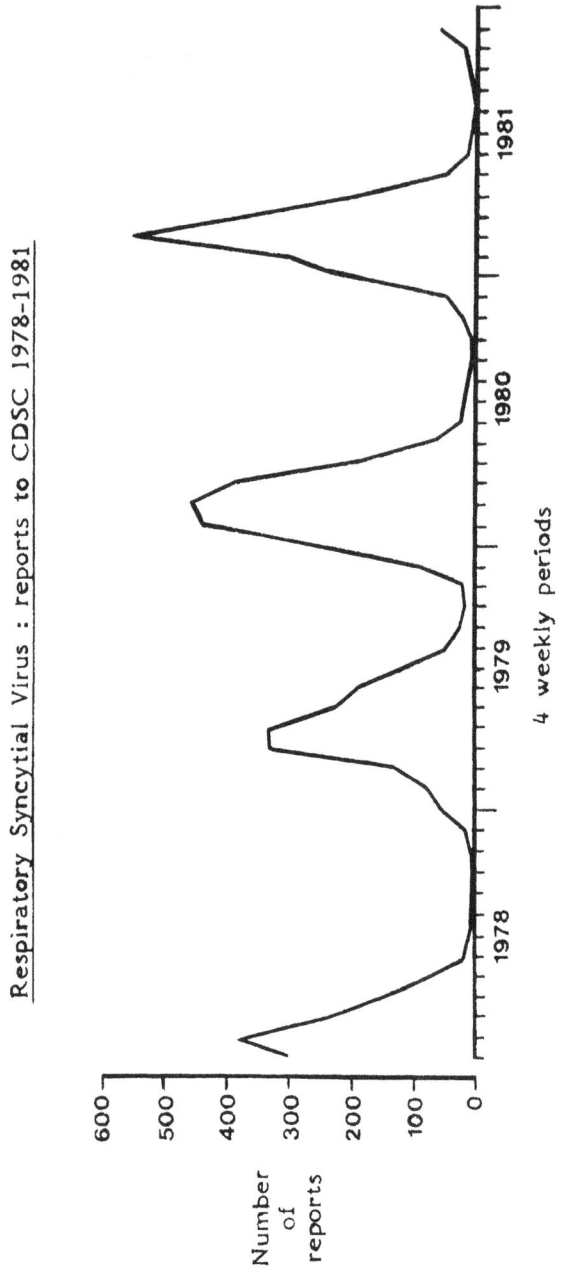

Fig. D 1.3.–RS virus over the period 1978-1981 (CDR 81/50).

Respiratory Syncytial Virus : reports to CDSC 1978-1981

4 weekly periods

Number
of
reports

as curves. The curves showing the incidence of colds week by week were quite complicated ones. The astonishing thing was that the complicated curves from one part of Holland could be fitted over those from another part of the country and the fit was remarkably close. This showed two things : first the time of rise and fall of the colds was almost exactly the same in different places, and second, the extent of the rise was also similar. He argued, not unreasonably, that all this would not fit in with a stepwise person-to-person spread. . . .Such findings are not islolated, very similar things have been reported by workers in the United States."

This result of Van Loghem proved that the cause of the colds, whatever it might be, was more or less contemporaneous over a fairly wide area, and observations in the United States have confirmed the result over areas still larger than the Netherlands. Sudden bad weather was at first thought to be a possible cause, people being weakened in some way by being physically cold or by being frequently wetted. Experiments under controlled conditions using volunteers have failed to provoke colds by such means, however, or indeed by any purely physical means. The remaining inference is that the common cold virus fell from the atmosphere over the geographical areas in question.

The common cold is one of a number of respiratory viral diseases which show peak incidences in W. Europe over the weeks from December to March. Thus Figure D1.1 shows a comparison of case numbers for Influenza A and Respiratory Syncytial (RS) disease over the winter of 1980–81, Figure D1. 2 shows Influenza B for the winter of 1981–82, and Figure D1.3 shows the situation for RS over the past few years. All this is strikingly similar to the annual breakdown through the stratosphere of small particles of viral dimensions revealed by studies of Rh-102 (entry A5). Similar **disease** patterns, but displaced six months, exist in the temperate regions of the southern hemisphere, and this too agrees with atmospheric behaviour in the stratosphere (c.f. R.E. Hope-Simpson, " The **influence of Season upon Type A Influenza** ", in *Biometeorological Survey*, Vol. 1, 170, Heyden, London). At the equator an indeterminate dependence of viral incidence on season is to be expected, and this is indeed seen in the data for influenza in Colombo, Sri Lanka shown in Fig. D1.4.

It is likely that clinical attacks of diseases are the exception rather than the rule. Most attacks are probably subclinical, not apparent to the eye, although on such occasions we tend to feel ' out of sorts ', without quite understanding why. Subclinical attacks are verifiable, however, by antibody and erythrocyte sedimentation techiques. Abnormally high rates affecting whole populations and probably covering a wide range of diseases have been reported (for example S. W. Tromp, *Experimentia*, 32, 1976, 126). Such occurrences may well be more common than one might think from a study of the literature, since we ourselves have received reports of this kind from hospitals by private communications, reports that so far as we are aware were not subsequently published. Widespread attacks on whole populations of many diseases more or less simultaneously are scarcely understandable except on the basis that the causative pathogens fell from the air

4 –

41

Average no. of Positive Sera at MRI from 1968 –1974

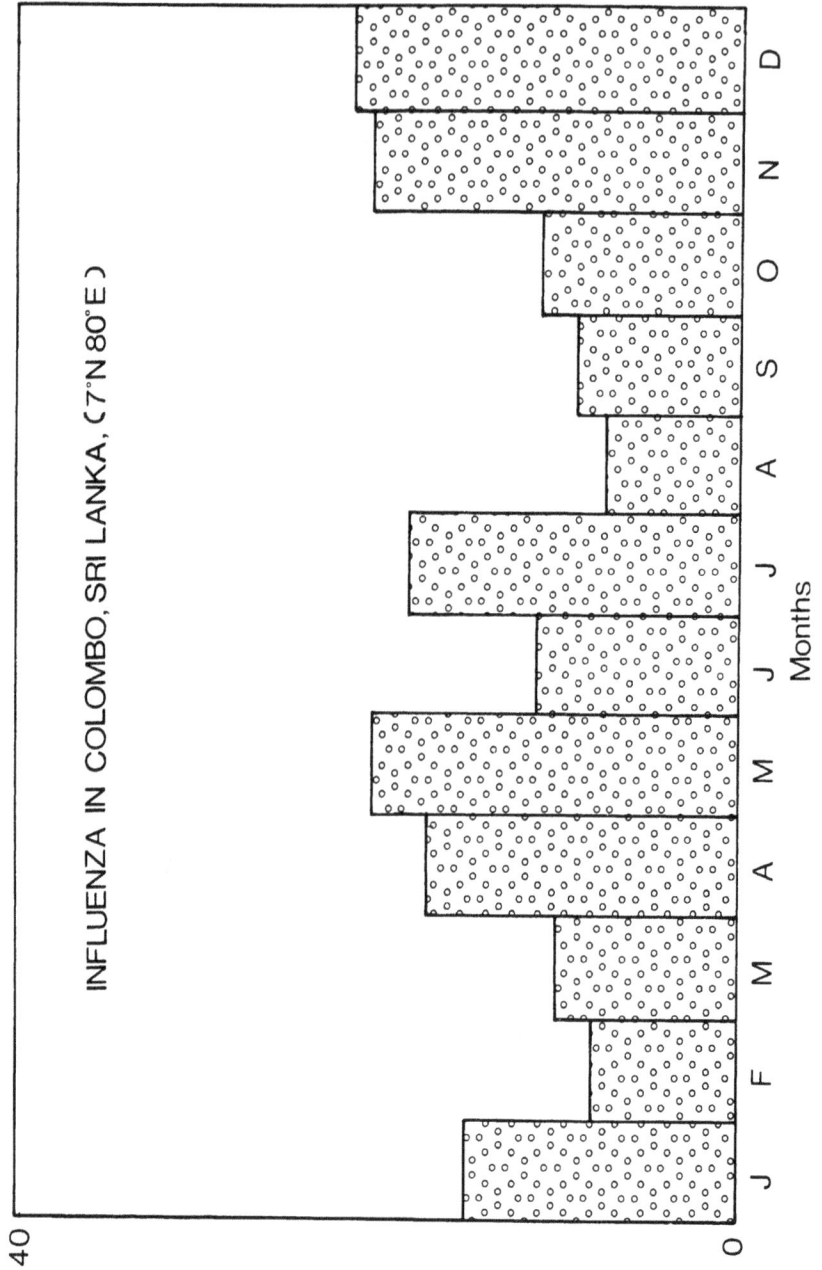

Fig. D 1.4.–The numbers of positive sera month by month over the period 1968-1974 for a uniform sampling of people in Colombo, Sri Lanka. Data supplied by the Medical Research Institute Virology Laboratory at Kalubowila, Sri Lanka.

INFLUENZA IN COLOMBO, SRI LANKA, (7°N 80°E)

Months

D 2. Some Dubious Claims for the Horizontal Transmission of Pathogens

Diseases in which a pathogen reaches a victim by transfer from another victim, whether of the same species or not, may be said to have been infected ' horizontally ', while a victim who acquires a pathogen falling through the air from high levels in the atmosphere (ultimately from space) may be said to have become infected ' vertically ' Since conventional theory holds that all diseases are in every case contracted horizontally, it follows that if some are in fact contracted vertically certain of the claims of conventional theory must be wrong. It is of interest to consider a few examples in which the usual claims are either dubious or certainly wrong (see also entry D4).

Table D2.1 gives data taken from the *Encylopaedia Brittanica* for the incidence of viral hepatitis in New York City, in suburban areas of New York State, and in the remoter regions of upstate New York.

TABLE D 2.1

Incidence of Viral Hepatitis

Region	Population Density (persons per acre)	Incidence (per 100,000)
City	24,000	10.4
Suburbs	515	19.2
Upstate	66	42.5

After expressing surprise at these data the author of the article in question makes the inference, having apparently never travelled on the New York City subway system, that people in country areas must be in closer contact with each other than people in the City. For any disease that is vertically incident, on the other hand, people who are most in the open air will be most at risk, and this is the natural interpretation of the data of Table D2.1.

In *J.A.M.A.*, Vol. 30, 1295-97, 1974, T. H. Lewis and W. L. Brannon discuss the curious case of the Trio Indians, a small tribe of about 500 persons formerly living near the equator in N. Brazil and S. Surinam. Tribes in this area had a reputation among explorers as ferocious bow-hunters, given to wiping-out any strangers who managed to penetrate the dense forests which existed before the present clearance schemes came into operation. By the early 1960's, however, the clearance scheme started by the Surinam Government brought the group of Trio Indians into the light of day. Exposed to the scrutiny of the world, survivors were found from attacks of poliomyelitis, which must have

43

occurred decades before the clearance scheme came into operation. Since such a small group, out of contact with the rest of the world, could hardly have sustained the polio virus over many centuries by internal circulation the strong indication is that the virus attacked the Trio by vertical incidence.

When many people hold a strong belief it is only too likely that they will not hesitate to use inconclusive data to build myths in its support. According to a widely-held belief, localised communities living in high latitudes soon become free of the common respiratory ailments if left to themselves, but they almost instantly succumb to such ailments on the arrival of visitors from the outside world. While the first part of this belief may have partial validity, it is certainly not strictly true. In the late months of **1918** isolated communities throughout the vast, lightly-populated State of Alaska were hit by devastating epidemics of influenza in circumstances in which travel was essentially impossible (entry D3).

The usual ' classic ' reference for thinking visitors so devastating to isolated communities is a paper by J. H. Paul and H. L. Freese (*American Journal of Hygiene,* 17, 1933, 517). Paul and Freese took up residence in a community of Norwegian miners and their families on the island of Spitzbergen towards the end of the summer of 1930, some nine months ahead of the crucial part of their investigation, which concerned the effect on the health of the community of the first spring boat of 1931, which arrived on 23 May in that year. From the beginning of September 1930 Paul and Freese took throat swabs from the people, which contributed nothing of relevance but which served to alert the population to the nature of the investigation being contemplated, an unwise step since it is a matter of experience that expectations feed on themselves. People were therefore will-primed when the all-important first spring boat of 1931 eventually arrived after the winter freeze-up to visit the surgery of the investigators. Indeed it was probably easy for the people to give the investigators exactly what they were seeking, lots of sneezing with the appearance of mild colds, since the miners and their families had lived through the winter in extremely crowded huts, doubtless overheated against the arctic winter, ideal conditions for extensive multiplication of the house mite in bedding materials. The first spring boat would arrive almost inevitably with the first tolerable spring weather, and with the first tolerable weather the womenfolk would be out shaking and beating the bedding vigorously, filling the air with house mite, one of the most ferocious of all allergic agents. Although the investigators persisted with their throat swabbings for almost a year, it does not seem to have occurred to them to watch-out for the house mite, likely enough the cause of many supposed outbreaks of the common cold in arctic communities.

The citizens of Charlottesville, Virginia, tended to suffer from colds during late September and early October. Someone noticed there was an excess of colds among children, which of course was natural enough because children have less well-developed, less ' experienced ', immunity systems than adults. Nevertheless, the myth became widespread that the children were transmitting the virus among themselves at school, school classes having started-up after the summer holiday. Nobody seemed to think it necessary to explain where the virus had come from in the first place, nor why a similar phenomenon did not occur everywhere throughout the world. The notion that the children first multiplied the virus at school and then spread it to their homes and thence through the city became widely believed.

It happened that a research group from the University of Virginia School of Medicine had the opportunity to put his superficially plausible notion to explicit test. J. O. Hendley, J. M. Gwaltney, Jr., and W. S. Jordan, Jr., remark in their paper in the *Amerian Journal of Epidemiology* (89, 1969, 184) :

> " . . . The peak occurrence of respiratory illness in the early fall occurred simultaneously in working adults both with and without exposure to children. In addition, the school children's illness rate peaked at the same time as adults. These epidemiological features suggest that variables other than school openings are important in triggering the fall outbreak of respiratory illness ".

The rashness with which those who believe solely in the horizontal transmission of diseases are ready to blame anything at all as a cause of epidemic outbreaks, provided only that it fits their prejudices, is well-illustrated by Figure D2.1. From this figure it is seen that the incidence of whooping cough has a clearly defined average periodicity of about three-and-a-half years. Advocates of horizontal transmission ' explained ' this periodicity in terms of a lag time required for the build-up of a new crop of susceptible children following each epidemic of the disease, an ' explanation ' that was almost obviously incorrect since each victim multiplies the whooping cough bacillus many millionfold, like a volcano bursting forth. To imagine that such explosive individual behaviour could be controlled by a weak annual variation of susceptibles, no more than a 25% effect, surely had to be wrong. Besides which, the same periodicity showed itself in city and rural populations alike, despite their greatly different population densities.

It became possible to test this theory because the population density of children susceptible to whooping cough was controlled during the period 1960-75, using a vaccine that conferred active immunity against the disease. During these years the susceptible pupulation fell to a far lower level than in earlier years. If the immunization had been 100% no disease would have occurred. In the actual situation immunization was not 100% across the population, but it was sufficient to lower the density of susceptibles so that according to the above theory epidemics should not have happened, or at any rate should have been more widely spaced than in the pre-immunization years. Yet as is seen from Figure D 2.1, the three-and-a-half year cycle continued unabated, but with fewer children involved in each **outbreak. The periodicity of whooping cough is reminiscent of** a pathogen derived from a cometary body with a viable halo spread appreciably along its orbit (entry C4).

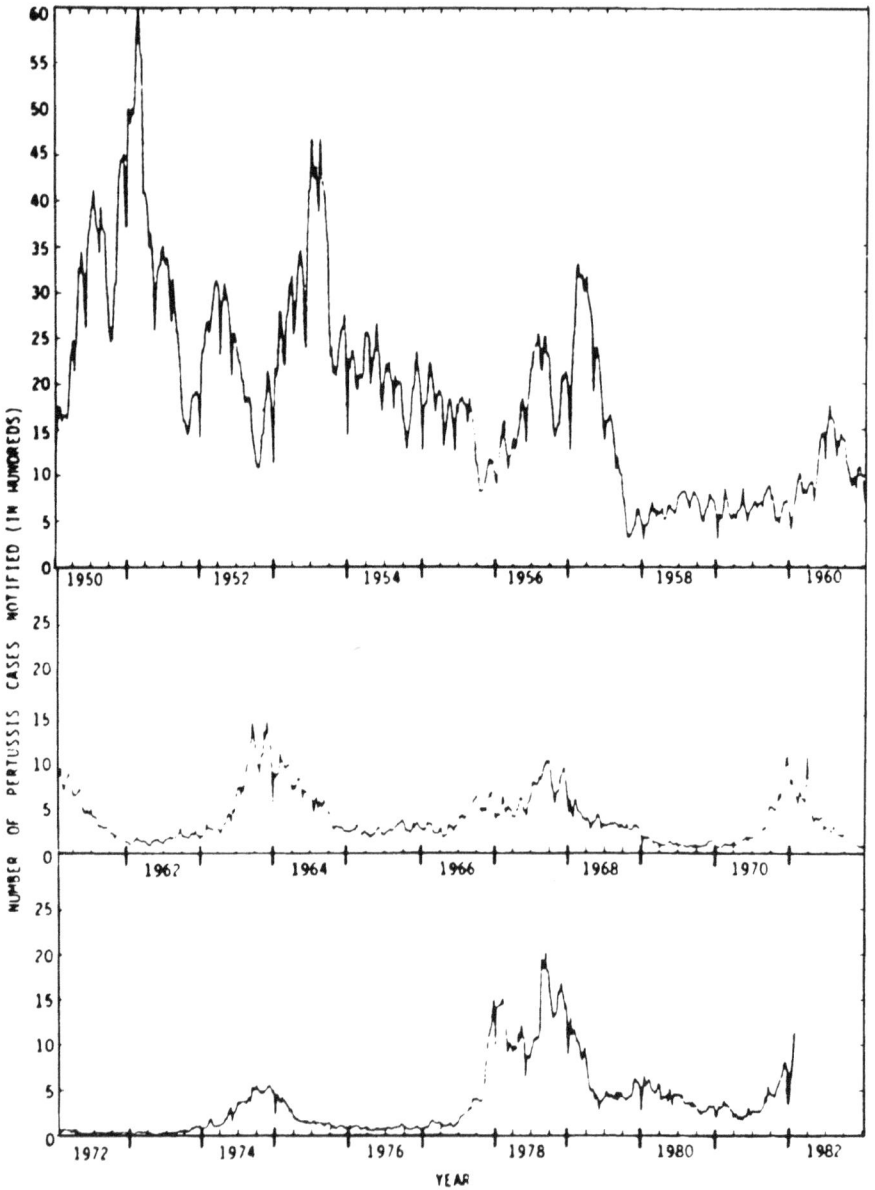

*Fig. D 2.6.–*Whooping cough notifications in England, Wales and Northern Ireland.

D 3. Evidence for the Vertical Incidence of Influenza A

The straightforward way to demonstrate the vertical incidence of a disease is to show that individuals out of contact with others nevertheless contract the disease. So it was for influenza in Sardinia in 1948. The detailed subtype of the influenza A virus dominant in 1948 first showed itself in Sardinia, an island where at that time communications were virtually non-existent. Reporting on his investigation of the outbreak, F. Magrassi (*Minerva med Torino,* 40,1949, 565) wrote :

> " We were able to verify . . . the appearance of influenza in shepherds who were living for a long time alone, in solitary open country far from any inhabited centre ; this occurred contemporaneously with the appearance of influenza in the nearest inhabited centres."

This observation shows that influenza can indeed be contracted without connection being necessary to another human. To explain Professor Magrassi's finding we must assume one of the following to have been true :

(a) The virus responsible for the Sardinia outbreak came from space. It happened for meteorological reasons (perhaps heat from the nearby active volcano of Stromboli) to touch down first in the area of the Tyrrhenian Sea.

(b) Winds blew the virus from some other area on the ground.

(c) Some animal exuded the virus all over Sardinia, and it passed to humans all in the same moment, whether in the remote countryside or in populated centres.

Of these, (b) is implausible because the new influenza varient started in Sardinia. There had been no other outbreak on the ground from which it could blow. As regards (c), there have been few reported cases of humans contracting influenza from animals. Influenza is said to have passed from pig to man at Fort Dix, New Jersey, in 1976. We mention this incident for those who wish to believe it. At all events, observed transfers from animal to man are so rare as to make it very doubtful that many such transfers could have occurred in Sardinia, and all at effectively the same moment.

Birds rather than pigs would be the least improbable animal to have produced such transfers. One might suppose that flocks of birds arrived in Sardinia from somewhere harbouring the new varient of the virus, and that by broadcasting bird muck all over the island they contrived to produce the remarkable contemporaneous outbreak investigated by Professor Magrassi.

In 1918 there were only some 45,000 people living in Alaska, which is two-and-a-quarter times larger than the State of Texas. In November and December of 1918 a lethal epidemic of influenza passed over the whole of that vast thinly populated territory, when human travel from the coast to the interior was essentially impossible because of snow and ice. Here again then we have an example of the spread of influenza by some means other than person-to-person contact, and as before there are the same three possibilities for explaining how the spread occurred.

In this case, it is possibility (c) that one must reject. Flocks of birds did not arrive in frozen November and December to spread ubiquitous muck all over the huge territory of Alaska. But the winter jet stream could quite well have overturned the upper atmospheric air, causing a cloud of virus-bearing particles to descend on even so large an area as Alaska. Indeed, it could have been the descent of bitterly cold air from the high atmosphere which caused Governor Riggs of Alaska to declare as follows to the Senate Committee of Appropriations (16 January 1919) :

" You have the short days, the hard cold weather, and you only make 20 to 30 miles a day over the unbroken trails. The conditions there are such as have never happened before in the history of the Territory ".

The influenza pandemic of 1918 had other peculiarities, Dr. Louis Weinstein wrote in the issue of the *New England Journal of Medicine* for May, 1976 :

" The influenza pandemic of 1918 occured in three waves. The first appeared in the winter and spring of 1917-18 . . . This wave was characterised by high attack rates (50 per cent of the world's population were affected) but by very low fatality rates . . . The lethal second wave, which started at Fort Devens in Ayer, Massachusetts, on September 12, 1918, involved almost the entire world over a very short time . . . Its epidemiologic behaviour was most unusual. Although person-to-person spread occurred in local areas, the disease appeared on the same day in widely seperated parts of the world on the one hand, but, on the other, took days to weeks to spread relatively short distances. It was detected in Boston and Bombay on the same day, but took three weeks before it reached New York City, despite the fact that there was considerable travel between the two cities ".

Here we have a third case where influenza was not spread by person-to-person contact, for nobody in 1918 could travel from Massachusetts to Bombay in the day or two which elapsed between the appearance of the new wave at Fort Devens and its appearance in Bombay. Nor could the fastest flying birds, the shearwater, swift, or even the albatross, make that journey in the time, even if Boston and Bombay were on the customary flight paths of these birds which they are not. Nor are there winds that blow over such a route, with such a speed, and over such a distance.

A markedly new variety of influenza falling from the high atmosphere, will in general arrive at ground-level at different places at different times. There will be a place where it arrives first,and this will be where the new epidemic begins, as in September 1918 at Ayer, Massachusetts. That the virus should happen to touch down shortly afterwards at two such widely- separated places as Boston and Bombay does not strain credulity at all. It was just that two of the first major patches of a virus which eventually descended on the whole world happened to hit Massachusetts on the one hand and the Bombay area on the other. Nothing is required to have travelled horizontally between these afflicted regions.

Since the atmosphere as it blows over obstacles on the land becomes a turbulent medium and since it is in the nature of turbulent motion that irregularities exist on all scales from the largest elements down to quite small dimensions, pathogens blown in the wind would be expected to reach ground-level in a distribution that was highly complex in its fine detail. In the early months of 1978 we obtained data on influenza outbreaks from more than 200 schools in England and Wales which proved the correctness of this picture. Table D 3.1 gives the absences from class of influenza victims, analysed with respect to house boarders at Headington School, Oxford. A small dense cloud of virus evidently fell on Latimer House over the weekend of February 4 and 5, demonstrating fine scale in the incidence of the virus over distances of   100 metres.

TABLE D3 . 1

Class-absences of Boarders at Headington School, Oxford
(classified according to Houses)

Date	Davenport (34 pupils)	Hillstow (63 pupils)	Latimer (46 pupils)	Napier (42 pupils)	Celia Marsh (44 pupils)
Jan 30	1	7	2	6	8
31	5	10	3	7	6
Feb 1	3	12	1	6	4
2	6	10	2	3	5
3	7	8	6	3	4
6	3	4	26	9	5
7	1	4	26	6	3
8	2	4	22	4	3
9	1	1	25	3	3
10	2	1	20	5	3
13	0	4	6	0	0
14	0	2	3	2	1
15	0	3	1	2	0
16	0	2	0	2	1
17	0	2	2	2	0

A viral particle is much too small to fall on its own accord under gravity from the tropopause to the ground. The particle falls because water condenses around it, until the aggregate becomes heavy enough to be pulled downward through the air. The condensate is initially ice, but the ice usually melts when the aggregate reaches the warmer air of the lower atmosphere, so giving a viral particle suspended in liquid water. Respiratory viruses arriving at the ground in a heavy shower of rain probably do little harm, since they are either washed away in streams or they adhere to grass, trees and other surface materials. However, there is a condition occurring at the end of a shower when falling droplets just fail to reach the ground, because the droplets evaporate some little way above the ground, thereby releasing viroids, viruses and other microorganisms into the air. The dangerous condition for infalling respiratory pathogens is therefore when it is *nearly raining*. Everyday observation shows that the phenomenon of nearly raining is very non-uniform over the ground, with some regions becoming wetted and other regions only a few yards away remaining dry. The fine detail of curtains of rain on a mildly showery day are very complex, a complexity well in accord with the highly variable incidence of influenza on the houses at Headington School, a variation that was far outside statistical expectation.

D 4. Evidence Against the Horizontal Transmission of Influenza A

Pupils in boarding schools do not stay segregated with respect to their houses during classes and games, and in many schools there is also a general mixing at mealtimes. If horizontal transmission of the influenza virus had occurred at all **extensively** during the incubation period of the disease, the intermingling of pupils would have precluded the large variation between Latimer House and the others shown in Table D 3.1. The virus would (with horizontal transmission) have become fairly uniformly spread across the school, which manifestly did not happen.

A phantom figure has been postulated for explaining the strange epidemiology of influenza, the illusory figure of the supershedder. A supershedder is supposed to act like a carrier of typhoid, an unidentified person who sheds the influenza virus in large quantity without apparently suffering from it, a person who cuts a swathe of disease through the unsuspecting surrounding population. But if a supershedder had existed in Latimer House, virus should have been exuded in school classes, where it would have become distributed approximately randomly with respect to the school houses. This would be true even if supershedders themselves succumb to the disease, since they would still exude the virus during incubation. Hence no school house could expect to escape infection by supershedders, regardless of which particular house the supershedders themselves were members. But school houses do escape infection, as for instance at Eton in 1978, where College House with 70 pupils had but a single victim in an epidemic with an average 35% attack rate for the whole school (441 victims out of 1248 pupils in the whole school). At another school where the two main houses each had about 55 pupils, one house had 2 victims and the other had 35 victims. Such cases, of which there are many, show the hypothesis of the supershedder to be wrong.

Fig. D' 4.1.

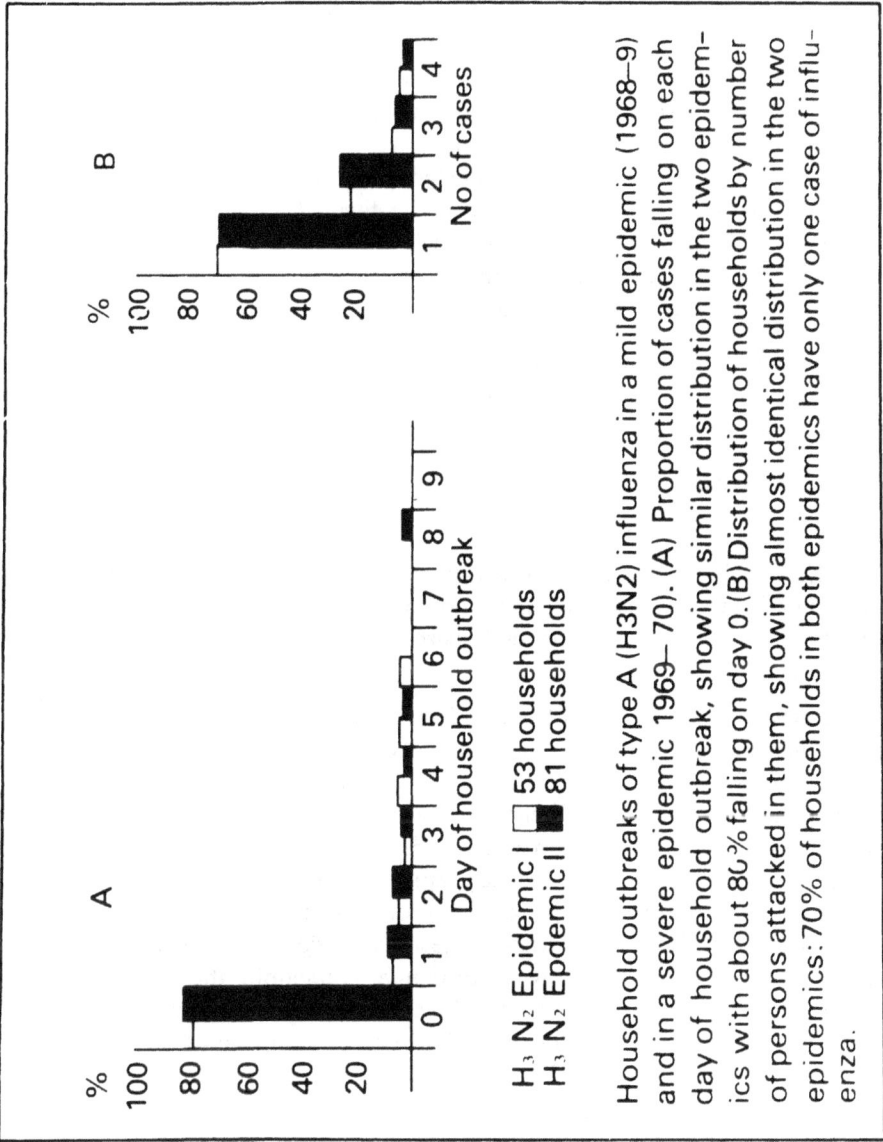

Household outbreaks of type A (H3N2) influenza in a mild epidemic (1968–9) and in a severe epidemic 1969–70). (A) Proportion of cases falling on each day of household outbreak, showing similar distribution in the two epidemics with about 80% falling on day 0. (B) Distribution of households by number of persons attacked in them, showing almost identical distribution in the two epidemics: 70% of households in both epidemics have only one case of influenza.

H₃ N₂ Epidemic I ☐ 53 households
H₃ N₂ Epdemic II ■ 81 households

One of the schools in our survey had conditions that were particularly well suited for testing the doctrine of case-to-case transmission. There were 48 victims among about 340 pupils, who all boarded under similar conditions. There were essentially 85 small dormitories, each with 4 pupils. The beds in each dormitory were close enough together to provide the readiest transfer of virus from pupil to pupil, if indeed infection goes at all in such a way. Thus case-to-case transmission would lead to the prediction of non-random clustering of the 48 victims in the 85 dormitories. There should be fewer dormitories with single victims in them than would be expected on a random basis, and more dormitories with clusters of two, three, and four victims in them. Calculation shows that a typical random situation would be 46 dormitories with no victims, 31 dormitories each with 1 victim, 7 dormitories each with 2 victims, 1 dormitory with 3 victims, and no dormitory with 4 victims. The actual results were as follows :

44 dormitories with no victims

35 dormitories each with 1 victim

5 dormitories each with 2 victims

1 dormitory with 3 victims

0 dormitories with 4 victims

The actual result was therefore very slightly less clustered than one would get in the most typical random distribution, but not by an amount with statistical significance. In effect, the situation was random, and there was no case-to-case transmission in the dormitories.

A very similar conclusion has also been reported by R. E. Hope-Simpson from a survey of influenza attack patterns in domestic households (*J. Hyg. Camb.*, 83,1979, 11). The histogram A of Figure D 4.1 shows the distribution of cases with respect to time, and histogram B shows the distribution of households by number of persons attacked in them, with open rectangles for 53 households in 1968-69 and solid rectangles for 81 households in 1969-70. It is immediately apparent from histogram A that even under living conditions of the greatest personal intimacy there could have been comparatively little transmission of the disease, while an easy calculation using histogram B shows that when there was more than one victim in a household the cases were mostly contemporaneous not sequential. It would be difficult to conceive of a more direct refutation of claims for the importance of horizontal transmission than these data.

D 5.   The Black Death

It is ironic that bubonic plague, the disease which historically caused the greatest devastations of human populations and which consequently inspired the greatest fear in man is not primarily a human disease at all. The bacterium *Pasteurella pestis* attacks many species of rodent, its main target. Because the black rat happened to live in close proximity to humans, nesting in the walls of houses, the physical separation of people from the black rat was small (especially for the poor living in tiny hovels) and could be bridged by fleas, which carried the bacterium from the blood of the rats to the blood of humans. This transfer was fortunately limited by the fleas themselves, since they preferred to stay with the rats, quitting them only as their hosts died from the disease. Although human-flea-human transfer of the bacterium presumably occurred, it does not seem to have been sufficient to maintain the disease, which died out as the supply of rats became exhausted.

The first clear reference to bubonic plague occurs in the *Bhagavata Purana,* an Indian medical treatise written in the fifth century B.C. There may have been an outbreak of plague during the first century AD, with the centres of the disease in Syria and North Africa, but between the first and sixth centuries there were no known outbreaks of the disease. However, in 540 AD a devastating pandemic involving the Near East, North Africa and Southern Europe is said to have had a death toll that reached 100 million, considerably more than the combined tolls of the two World Wars of the present century. This was the so-called Plague of Justinian.

Bubonic plague would seem then to have disappeared for eight centuries until the shattering outbreak of the Black Death in 1347–50. Thereafter, the disease smouldered with minor outbreaks until the mid-seventeenth century. One learns in history that the disappearance of plague in the seventeenth century was due to the improving construction of houses, which afforded less opportunity for the nesting of the black rat. The Great Fire in 1666 is said to have put an end to the disease in London, by burning-out all the old hovels. It is a sad comment on the value of education that probably ninety-nine out of a hundred students believe this facile story, without pausing to reflect that it gives no explanation of the eight-centuries remission of the disease from 540 to 1347–50, or of its decline following the seventeenth century in the rest of the world, where plenty of hovels still remained, or of its recent emergence in China in 1894. In India, it killed some thirteen million people in the years up to the first World War.

In entry C4, the encounters of the Earth with a compact halo of viable microorganisms issuing from a cometary body is considered. For an explicit case in entry C 4, the pattern beginning in the year $t = 0$, is for encounters in the following particular years :

$$t = 0, 114, 228, 342, 456, 1151, 1265, 1379, \ldots$$

Another case gave encounters in the years

$$t = 0, 233, 276, 509, 552, 785, 1061, 1294, 1337, \ldots$$

The pattern is of a number of encounters about a century or two apart followed by a considerably bigger jump, analogous to the historic situatuation for bubonic plague.

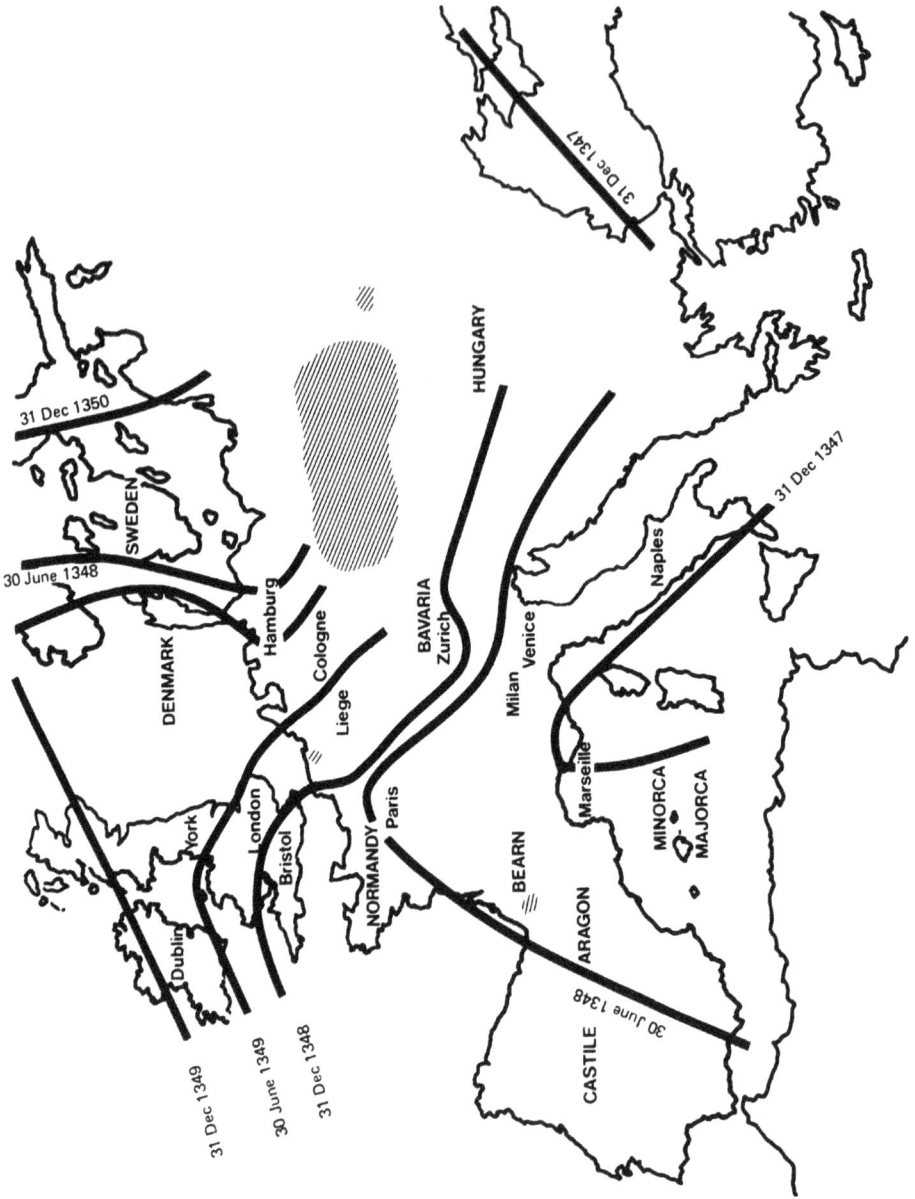

Fig. D 5.1   Contours of Black Death.

By working from historical records of the first outbreaks of the Black Death in various population centres, Dr. E. Carpentier obtained the results shown in Figure D5.1, where the contours are drawn so as to connect places first reached by the disease at the dates, marked on the contours. The contours are not of equal reliability. Because of the widespread availability of reasonably accurate records in W. Europe the contours in the western part of the figure must be quite reliable, whereas the contour of 31 December 1350 for the eastern Baltic is surely more uncertain. The least reliable contour, the segment from the Crimea to the Bosphorus, comes not from archives but from a story. Whenever a new disease breaks out, people always look for a scapegoat, for some other place (dirty foreigners) from which the disease has come. Even Thucydides, 'the father of history', fell into this emotive trap, by ascribing the origin of the pathogen responsible for the Plague of Athens to the distant 'Libyans'. Outbreaks of foot-and-mouth disease among farm animals in modern times are always supposed to come from the carelessness of farmers somewhere else. A new canine viral disease appeared a year or two ago, essentially simultaneously all over the world. This was blamed variously on cats, on judges travelling from one dog-show to another, and on a supposed faulty issue of vaccine from some unspecified laboratory. A new venereal disease also appeared a year or two ago, among horses. This was not much mentioned in the media, because of the widely-held belief that venereal diseases are the just deserts of sexual promiscuity and it was hard to attribute much in the way of licentious behaviour to horses. The affair caused a great deal of ill-feeling internationally among owners of valuable racehorses, who were all convicced, every manjack, that the fault (the villany even) lay with the other fellows. In a similar way the Europe of the Black Death blamed the Middle and Far East. The disease was supposed to have been brought into Europe from the Tarters who had beseiged the Genoese base of Caffa in the Crimea. The dubious contour for 31 December 1347 is said to mark the voyage of Genoese ships back to Italy.

In entry A4 it was shown that pathogens of bacterial size take a year or two longer to fall under gravity through the stratosphere at high latitudes than they do in equatorial regions. This is because the tropopause is more than 5 km higher at the equator, so that pathogens have less still air to fall through there than they have at higher latitudes. If the land surface were everywhere low and flat, contours giving equal times of arrival at ground-level of a bacterium injected everywhere at essentially the same time at the top of the stratosphere would run generally east-west, and with increasing delay in arrival times occurring from south-to-north. High mountains like the Alps would, however, introduce perturbations in such a simple picture, since high mountains have altitudes that are a significant fraction of the height of the tropopause. Looking at Figure D 5.1 from this point of view, the bulge in the contours caused by the Alps is clearly apparent. With this bulge allowed for, the contours do run generally east-west, and there is indeed a delay from south-to-north. For a bacillus of the size of *Pasteurella pestis* (rod-diameter about 0.7 μm) the delay from latitude 35° to latitude 60° should be about 2 years (entry A 4) which is just what Figure D 5.1 shows.

55

It is worth considering the hopeless tangle into which the contours of Figure D 5.1 inevitably lead one according to conventional theory, according to which they are to be interpreted as steps in the march of an army of plague-infested rats, a march which began at the end of December 1347 at the port of Genoa. Humans with the disease collapsed on the spot and afflicted rats must surely have done the same–the *Bhagavata Purana* explicitly advises people to leave houses and other buildings ' when rats fall from the roofs above, jump about and die '. How close the writers of this ancient treatise were to the truth, already in 500 B.C. ! To argue that stricken rats set out on a journey that took them in six months, not merely from southern to northern France, but even across the Alpine massif, borders on the ridiculous.

Nor does the evidence by any means support in detail such an inexorable stepwise advance of the plague. There were lots of local irregularities such as is described for influenza in entry D 3. Pedro Carbonell was the archivist to the Court of Aragon, a post which in its nature could only be held by a person with a clear appreciation of the difference between fact and fiction. Carbonell reports that the Black Death began in Aragon, not at the Mediterranean **coast** or at the eastern frontier, as it should have done if there had been a marching army of rats, or rats carried in ships, but at the inland city of Teruel.

Since it apparently stretches credibility too far to argue that the advancing army of stricken rats also managed to swim the English Channel, it is said that the Black Death reached England by ship. Yet the contours of Figure D 5.1 are of quite the wrong shape for boats to have played a significant role in spreading the disease. If *Pasteurella pestis* had been spread by sea, the earliest contour would be wrapped around the coastline from the Mediterranean to northern Europe, with subsequent contours then filling-in gradually towards central Europe. Not only this, but if the bacillus had travelled by sea, the coast of Portugal would have been seriously affected, whereas the evidence is that the plague scarcely penetrated to Castile, Galicia and Portugal.

There are many descriptions of communities which isolated themselves deliberately from the outside world, many such descriptions from English villages. Yet isolation was to no avail. The phantom figure that brought the disease always managed to evade the precautions. The Black Death would strike suddenly, and within a week an isolated community, the remotest village, would be just as affected by the disease as the City of London. What remarkable rats they must have been, to have made a beeline for remote villages and yet to have spared the shaded areas of Figure D 5.1, to have spared even the population centres of Milan, Liege and Nuremberg ! The astonishing reason offered for the good fortune of the extended area of Bohemia and southern Poland is that the rats disliked the food available to them in these regions.

Unless one is irreversibly programmed by modern **superstitions**, the contours of Figure D 5.1 are a clear indication that *Pasteurella pestis* hit Europe from the air. There was no marching army of plague-stricken rats. The rats died in the places where they were infected, just as humans did. By falling from the air *Pasteurella pestis* had no difficulty in crossing the Alps, or in crossing the English Channel. Remote English villages were hit, however determinedly they sought to seal themselves off from the outside world, because the plague bacillus descended upon them from above, and against an aerial assault they had no remedy. Milan, Liege and Nuremberg went comparatively unscathed because it is in the nature of atmospheric turbulence (entry D 3) that there will be odd spots where a pathogen does not fall. So too did Bohemia and Southern Poland escape, even though these regions grow food just as palatable to rats as everywhere else.

D 6.   Comets and Superstitions

According to entry C2, smaller cometary bodies with masses of order 10^6 tons are likely to be more important sources of terrestrially-incident microorganisms than full-blown comets with masses of order 10^{10} tons. This is because the smaller bodies are probably much more numerous, so that the Earth is far more likely to come close in a specified interval of time, say a year, to a smaller body than to a full-blown comet. The distance a microorganism must travel to reach the Earth when its parent body passes closeby is comparatively small, and the exposure to harmful radiation from the Sun (entries B2 and B3) is consequently much less than for passages at distances of the order of the radius of the Earth's orbit around the Sun—the usual situation for full-blown comets. Thus large seemingly spectacular comets are likely to be of less consequence on a regular basis for the supply of viable microorganisms to the Earth than are the smaller bodies considered in entry C2, smaller bodies which it would be difficult to observe visually—nobody had any warning of the body which hit Siberia in the region of the Tunguska River in 1908.

Yet if one waits a long—enough time, a full-blown comet will sooner or later pass the Earth sufficiently close for viable microorganisms evaporated from the comet to reach the safety of the terrestrial atmosphere. When this happens, because of the largeness of the comet, the supply of microorganisms would be exceptionally great, and pathogenic attacks on terrestrial plants and animals would be particularly severe. The widespread incidence of epidemics of all kinds following closely on the appearance in the sky of an exceptionally brilliant full-blown comet (exceptionlly brilliant because of its closeness) would not be lost on the human population. Nobody actually living through such a holocaust of disease would have any doubt about its origin in the visitation of the comet. So the subsequent appearance of any comet at all in the sky would be an occasion for fear. However, as the generations passed and the experience was not repeated because of its rarity, 'enlightened' individuals would come eventually to declare the old belief a superstition. For a while there would be a clash between the old believers and the

5-

enlightened ones, but as each further comet (being normally distant from the Earth) brought no ill-effects in its train the clash would subside with the population coming to accept the supposedly enlightened view–until the next close approach of a full-blown comet occcured, when the old belief would again be seen to be correct.

Was this the rationale behind the medieval dread of comets it is natural to ask ? The question is perhaps best addressed by visiting a medieval cathedral. Consider the determination and ingenuity that must have been needed to erect such a structure, equipped only with primitive tools. Consider where the stone came from. At Ely, for instance, the builders had the forethought to bring it from a considerable distance in small boats, a hard limestone that gave a thoroughly durable structure. Were these builders and their workmen the kind of people to be terrified by an insubstantial superstition ? People in medieval times had plenty of really serious troubles to worry about (with no dentists around think how much toothache there must have been, a mere detail no doubt, but think of it nevertheless !). Is it not our own supposedly enlightened age which has the comfort to permit itself to be deceived by **superstitions**?

D 7.   The History of Diseases

If for the reasons developed in entries C2 and D6 most pathogens are derived from cometary bodies with masses of 10^4 to 10^6 tons rather than from full-blown comets with masses of order 10^{10} tons, the supply of pathogens from a particular body cannot last for more than about a hundred perihelion passages (entry C5). If all cometary bodies contained exactly the same microorganisms the cessation of one body as a source of pathogens would not be noticed so long as the demise of the body was compensated by the addition from the outer regions of the solar system of a new body of comparable size which had experienced a suitable shortening of its orbital period due to gravitational perturbations (entry C4). But it seems unlikely that microorganisms in one body would be precisely the same as in other bodies. Replications within each body would generate variations, even if microorganisms in different bodies had an initially common source. Such variations for pathogens would generate similar but not precisely the same diseases of plants and animals. Thus in the several centuries, or the one or two millennia, needed for a set of cometary bodies to be evaporated and to be replaced by another set there would be noticeable changes in the general pattern of diseases. Not every disease need change of course, because some may be maintained more or less steadily by horizontal transmission (e.g. tuberculosis, herpes simplex) while others may be derived from much larger cometary bodies, say 10^8 tons, with correspondingly longer lifetimes as suppliers of pathogens. Nevertheless, the expectation is of a pattern of diseases subtly changing down the centuries, and the question to be considered here is whether the history of diseases supports this point of view.

Some diseases present in recent times were almost certainly absent in ancient times, and some that were virulent in ancient times have been absent in modern times. To consider examples, smallpox and whooping cough were almost surely absent in ancient times, while the plague of Thucydides (entry D8) and the English Sweats of the late Middle Ages have not been seen in modern medical practice. The absence of smallpox is shown by the lack of a word for pockmark in either ancient Greek or Latin, by the absence of a description of smallpox in the Hippocratic Writings, and by the lack of mention throughout classical literature of persons with faces scarred by smallpox. It is remarkable that no clearly recognisable description of whooping cough appears to be known before the end of the sixteenth century A.D. A young child with whooping cough experiences symptoms that are both alarming and easily described. Inspiration stops for a long as fifteen normal breaths, the child begins to go blue in the face, and then—just as the parent thinks the child is choking to death—there comes the furious intake of air which causes the ' whoop ' It seems inconceivable, if this disease existed in Greek and Roman times, that contemporary medical treaties would have avoided mentioning it.

The Hippocratic Writings must make somewhat frustrating reading for the professional physician. While a handful of diseases are discribed with clarity, the general impression is of a situation out-of-focus, of writers who did not quite have a firm grip on their subject matter. For example, the writers used the word *causus* for some kind of fever without bothering with a precise definition of it, as if everybody knew *causus* when they saw it. No modern translator has been able to identify *causus*, likely enough because no modern physician has ever seen the disease. Because translators have worked on the assumption that Greek diseases were the same as ours, they have been obliged to suppose the Greeks confused together in *causus* a whole lot of sources of fever, and unlikely supposition.

There is a disposition to think that because people in earlier times understood less than we do today their eye-witness descriptions of phenomena were inferior to ours. Precisely the reverse is almost surely true. When people are aware of their ignorance they take an intense pride in accurate description, since this is the one activity available to them which can be done well. When people know a little their descriptions tend to be distorted to correspond to what they believe to be true, and when people know a lot they give more attention to thinking and less to description, and so are led into some inaccuracies due to oversights. We therefore believe that the Hippocratic Writings were accurate descriptions, not of modern diseases to be sure, but of diseases as they existed in Greek times, diseases with family resemblances to ours but subtly different.

59

D 8. Plague at Athens

A virulent disease broke out in Athens in 430 BC. The Pelopponesian War, which marked the decline of Athens, had begun the previous year. The history of the War down to 411 BC was written by Thucydides with what has been called 'minute and scientific accuracy'. He describes the 'plague', evidently not bubonic plague, as follows :

"The season was universally admitted to have been remarkably free from other sicknesses ; and if anybody was already ill of any other disease, it finally turned into this. The other victims who were in perfect health, all in a moment and without any exciting cause, were seized first with violent heats in the head and with redness and burning of the eyes. Internally, the throat and the tongue at once became blood-red and the breath abornmal and fetid. Sneezing and hoarseness followed ; in a short time the disorder, accompanied by a violent cough, reached the chest. And whenever it settled in the heart, it upset it, and there were all the vomits of bile to which physicians have ever given names, and they were accompanied by great distress. An ineffectual retching, producing violent convulsions, attacked most of the sufferers ; some as soon as the previous symptoms had abated, others, not until long afterwards. The body externally was not so very hot to the touch, not yellowish but flushed and livid and breaking out in blisters and ulcers. But the internal fever was intense ; the sufferers could not bear to have on them even the lightest linen garment ; they insisted on being naked, and there was nothing which they longed for more eagerly than to throw themselves into cold water ; many of those who had no one to look after them actually plunged into the cisterns. They were tormented by unceasing thirst, which was not in the least assuaged whether they drank much or little. They could find no way of resting, and sleeplessness attacked them throughout. While the disease was at its height, the body, instead of wasting away, held out amid these sufferings unexpectedly. Thus, most died on the seventh or ninth day of internal fever, though their strength was not exhausted ; or if they survived, then the disease descended into the bowels and there produced violent lesions, at the same time diarrhoea set in which was uniformly fluid, and at a later stage caused exhaustion, and this finally carried them off with few exceptions. For the disorder which had orignally settled in the head passed gradually through the whole body and, if a person got over the worst, would often seize the extremities and leave its mark, attacking the privy parts, fingers and toes ; and many escaped with the loss of these, some with the loss of their eyes. Some again had no sooner recovered than they were seized with a total loss of memory and knew neither themselves nor their friends."

This passage provides a serious challenge to anyone who believes in even an approximation to a historical constancy of diseases. Without reading anything further of Thucydides, it is obvious that here is a writer in total command of his craft, able to express himself in great detail and yet with a remarkable economy of words. Thucydides saw the disease all around him, the description is a thoroughly first-hand account, the information is generous in quantity, and yet no   medical authority has succeeded in identifying the disease. And not for want of trying, for as one commentator remarked :

" I have looked into many professional accounts of this famous plague, and writers, almost without exception, praise Thucydides' accuracy and precision, and yet differ most strongly in the conclusions they draw from the words. Physicians—English, French, German—after examining the symptoms, have decided it was each of the following : typhus, scarlet, putrid, yellow, camp, hospital, jail fever ; scarlatina maligna ; the Black Death ; erysipelas ; smallpox ; the oriental plague ; some wholly extinct form of disease. Each succeeding writer at least throws doubt on his predecessor's diagnosis ".

The game is one that anyone can play. Given a modern text on infectious diseases, find a correspondence with Thucydides. We tried ourselves, beginning by casting a wide net, trying many possibilities. Our conclusion was that smallpox in its dangerous confluent form had the most points of similarity with Thucydides. Headache, high temperature, smell, hoarseness, ineffectual retching, and thirst, are all found in confluent smallpox. The words ' breaking out in blisters and ulcers ' implies pustules in the skin rather than a rash of spots on the skin, and this too is correct. Most diagonstic is the passage : ' While the disease was at its height. . . finally carried them off with few exceptions '. Thus in a modern text :

> " In fatal cases by the tenth or eleventh day the pulse gets feebler and more rapid, the delirium is marked, there is sometimes diarrhoea, and with these symptoms the patient dies "

It is also correct that eyes could be ' lost ' through subsequent blindness and that fingers and toes could be lost through gangrene, although these complications are apparently not common in modern times. Post-febrile insanity is sometimes met, which could conceivably explain the last sentence of the quotation : " Some again had no sooner recovered. . . "

Correct further statements appear in later paragraphs of Thucydides. Thus :

> " Equally appalling was the fact that men died like sheep, catching the infection if they attended on one another ; and this was the principal cause of mortality " . . . . . .

> " For no one was ever attacked a second time, or not with fatal result".

We were not overwhelmed by these successes, however - there were troubles that will be mentioned in a moment — but the successes seemed sufficient to take our investigation to the point of determining more exactly the meaning in several places of the original Greek. The phrase rendered in the above translation due to Ben Jowett as ' most died on the seventh or ninth day of internal fever ' has always caused difficulty. Faced with the improbability that death somehow skipped the eighth day, some translators have taken the liberty of changing nine to eight, thus presuming that the father of history was unable to count ! Dr. Humphrey Palmer of University College, Cardiff, informed us that a literal translation reads 'most died as niners and / or seveners', a niner being a colloquialism for a victim who died in nine days, and a sevener for one who died in seven days. Thus the usual rendering (as in Ben Jowett's translation) is consistent with the original, but if this was what Thucydides intended one has to wonder why a writer of such acknowledged precision shoud have inverted the usual order of seven and nine. In the days of Thucydides there was a general confusion throughout the Mediterranean, as to whether weeks should be counted in seven days or nine days. Even two centuries later the weekly markets in Rome were held at nine-day intervals. This raised the question of whether the inverted order 'niner *and* sevener', was a colloquialism for a fortnight, which would be helpful to confluent smallpox.

If one were to accept the latter interpretation, two difficulties for smallpox still remained. All modern descriptions speak of victims complaining in the early stages of severe lumbar pains, whereas Thucydides is insistent : 'For the disorder which had originally settled in the head . . . . . . ' There is no mention of pains in the back and Thucydides should certainly have known since he was himself a victim who managed to survive the disease. More important, there is no mention of the very high **density** of pustules on the face that is the main outward visual characteristic of confluent smallpox. Nor is there any mention of the scarring and disfigurement of survivors. It hardly seems credible that an accurate observer would by-pass this most terrrible aspect of confluent smallpox. However hopeful one is in the beginning of making a postive identification, whatever choice of disease one makes, the trial always seems to peter out in this way. There were evidently aspects to the Plague at Athens that do not fit any modern disease, as the above quotation, 'I have looked into many **professional** accounts . . . . . ' already implied.

E

Evolution (biological)

E 1. Evolution, a Brief History of the Darwinian Theory

It is obvious to the eye that remarkable similarities exist between animals and plants which yet do not normally interbreed with each other, between related species as one says, and this fact must have been known for thousands of years. When the idea first suggested itself to some person that apparently related species really had been related in the sense of being derived from a common ancestral species is not known, although towards the end of the seventeenth century Robert Hooke, who coined the word 'cell' used so widely in modern biology, is said to have been of this opinion. By the latter half of the eighteenth century the evolutonary view had become widespread, particularly in France, to a degree where the systematist Linnaeus accepted it around the year 1770 in order it seems to avoid being castigated by his contemporaries as a fuddy-duddy.

The first widely-discussed evolutionary theory was published in 1809 under the title *Philosophie Zoologique* by J-B de M. Lamarck. The theory rested on the postulate that special characteristics acquired by struggles for existence during the lives of parents tend to be transmitted to their offspring. If this postulate had been true, the theory itself would have been logically viable, but many subsequent experiments have shown Lamarck's axiom to be wrong, unfortunately for him.

British naturalists did not begin in the first third of the nineteenth century with a view as wide as the French had held in the eighteenth century, perhaps because of a distrust in Britain, following the Revolution of 1791 - 94 and the Napoleonic Wars, of everything French. The initial concern of British naturalists was to understand the factors in nature which control the balance of the varieties of a single species. Since the varieties could be observed actually to exist, they were accepted as given entities, requiring no explanation, thus avoiding the pitfall of Lamarck.

It has been said that the first mention of natural selection was made by William Wells at a meeting of the Royal Society of London as early as the second decade of the nineteenth century. The phrase 'natural process of selection ' was explicitly coined by Patrick Matthew in *Naval Timber and Arboriculture* published in 1831 (Edinburgh). The idea of natural selection is really no more than a tautology :

If among the varieties of a species there is one better able to survive in the natural environment that particular variety will be one which best survives. The powers of invention required to perceive this truism could not have been very great.

If evolution leading to the divergence of species from a common ancestor was suspected, and if the concept of natural selection was available, why was the theory of evolution of species by natural selection not under discussion already in the 1830's ? The answer is that it was, as can be seen from the second of two papers pubished in 1835 and 1837 by Edward Blyth (*The Magazine of Natural History*). The first of these papers, *The Varieties of Animals,* is a classic. Besides the clarity with which Blyth addressed his main topic the paper contains passages which foreshadow the later work of Gregor Mendel. In his second paper, Blyth considered the theory of evolution of species by natural selection, telling us in passing that the matter had frequently been dealt with by abler pens than his own. The difficulty for Blyth was that, if 'erratic adaptive changes' as he called the modern concept of mutations could arise spontaneously in a species, why were species so sharply defined ? Why was the common jay so invariant over the large latitude range from S. Italy to Lapland, when surely it would be advantgeous for appreciable variations of the jay to have developed in order to cope better with such large fluctuations in its environment ? So quite apart from the unsolved question of the source of the supposed mutations it seemed to Blyth as if the evidence did not support the concept of evolution by natural selection.

The position remained unchanged in this respect for two further decades until the arrival of a new generation of British naturalists, a position analogous to that which occurred almost exactly a century later in respect of the theory of continental drift. In spite of there being evidence in favour of continental drift, geologists and geophysicists convinced themselves in the 1930's that there were overriding reasons why the theory could not be correct, However, the evidence continued to accumulate to such a degree that by 1960 the situation became inverted. The evidence forced scientific opinion to accept the theory of continental drift, even though nobody understood why continents drifted. So it was with the theory of evolution by natural selection. The evidence forced belief in the theory, even though nobody understood why mutations occur or how the difficulties raised by Edward Blyth might be overcome.

The two crucial papers were both written by Alfred Russel Wallace, with titles that left little doubt of their author's intentions, in 1856 *On the Law which has Regulated New Species,* and in 1858 *On the Tendency of Varieties to Depart Identifinitely from the Original Type.* Unfortunately for **Wallace** and for scientific history, he chose to send both papers to Charles Darwin, who had himself been skirting the problem for many years in his personal writings, but who had published nothing nor even communicated his views to his closest friends. With Wallace's second paper available to him however, Darwin then wrote his book *The Origin of Species* published in 1859. The surprise is that, in spite of the extreme clarity of Wallace's writing, Darwin still contrived to state the theory in a laborious confused way and with an erroneous Lamarckian explanation for the origin of mutations, an explanation which Wallace had himself explicitly eschewed (for a detailed discussion see C. D. Darlington, Darwin's Place in History (Oxford, 1959)).

If Wallace had published his papers **quietly** in the Journal of the Linnaean Society his views would probably have made as little immediate impact as did the now-classic paper of Gregor Mendel. It was the social prestige enjoyed by Darwin, his friends and supporters, that brought the theory of evolution by natural selection forcibly on the world's attention. As always seems to happen when media publicity becomes involved nobody was then interested in precise statements or in historic fact. Writers copied from each other instead of checking original sources, careers were based on the controversy, and attributions became falsified. So did it come about that the theory became known as Darwin's theory, just as two decades earlier the ice-age theory had become known as Agassiz' theory, after Louis Agassiz who propagandised effectively for that theory but did not invent it.

65

E 2. The Neo-Darwinians

The work of Gregor Mendel (published in 1866), was rediscovered early in the present century. The work showed that certain heritable characteristics, colours of peas in Mendel's case, were determined by a discrete **unit**, which was transmitted from generation to generation in accordance with certain simple mathematical rules. Generalising from the small number of characteristics involved in the early experiments, the view soon gained ground that all the gross characteristics of a plant or animal were determined by small discrete units, genes. At the suggestion of W. Johannsen in 1909, the inferred collection of genes for a set of identical individuals in a species became known as their genotype, and the plant or animal to which the geneotype gave rise was called the phenotype.

Advances in microscopy pointed to certain discrete objects in the nuclear region of cells, the chromosomes, as the likely site of the geneotype. Since the inferred number of genes was much greater than the number of chromosomes, the genes became thought of (correctly as it eventually turned out) as small structures carried on the **chromosomes.** Microscopy was not sufficiently refined, however, for individual genes to be distinguished, only the gross forms of the chromosomes. The gross forms for a particular organism became know as its karyotype. Grossly different organisms had readily distinguishable karyotypes, but similar species were often found to have karyotypes that could not be distinguished by the microscopic techniques then available. It was felt, however, that a **detailed knowledge of the genes–if it were available–would distinguish between similar** species, or even between varieties of the same species. How far this has turned out to be true will be considered in entry E6.

Experiments of genetic significance in the first half of the century were mainly of two kinds, more complicated examples of the cross-breeding of varieties than those examined by earlier workers, and experiments designed to induce changes in the genotype. Since a gene is a material structure, it was argued, the structure must be changeable by violent means, through irradiation by X–rays for example. It was found possible in some cases to induce changes by such means without destroying viability, although for the great majority of changes viability was weakened in comparison with the original organisms. So genes could be changed, organisms could be altered, mutations could happen it was proved, even though the mutations were deleterious in the overwhelming majority of cases.

Since there could be mutagenic agents in the natural environment, for example the near-ultraviolet component of sunlight and ionizing radiation from cosmic rays, mutations could arise in the wild. Besides which, it is surely impossible to keep on copying any object or structure without an occasional error being made. So quite apart from deliberate **mutagenic agents there must be an non-zero copying error rate occurring in the genotype** from generation to generation. Here at last therefore were the mutations required by the

Fig. E 2.1. Enzyme action : formation of an enzyme-substrate complex, followed by catalysis.

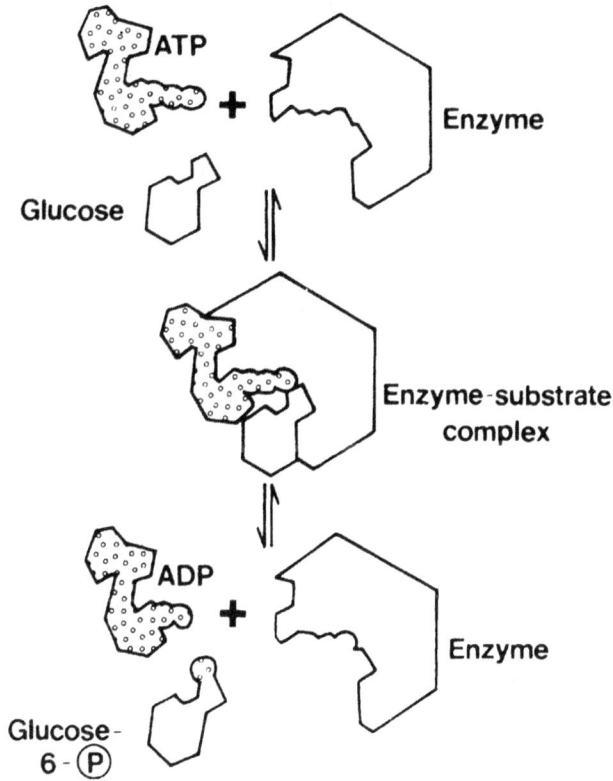

theory of evolution through natural selection. No matter that most of the mutations would be bad, since the bad ones could be removed by natural selection it was argued (erroneously as will be seen in entry E5). Such then was the position of the neo-Darwinians, who imagined themselves in a stronger position than the biologists of the nineteenth century had been, but the reverse was actually the case. The theory in the form proposed by Wallace would admit of mutational changes coming from anywhere, by additions to the genotype of a species from outside itself, for example through the addition of externally incident genes, as well as by changes to already-existing genes. The neo-Darwinians were confined, however, to the already-existing genes, and this had turned out to be an insufficient position, as will be demonstrated here and in entries E4 and E6. The neo-Darwinians boxed themselves into a closed situation, whereas the theory of Wallace could be either closed or open.

The development of modern microbiology from the work of Oswald Avery in the mid-1940's, through that of Erwin Chargaff to the elucidation of the structure of DNA by Francis Crick and James D. Watson, added precision to the concept of the genotype. The genes were sequences of four kinds of base-pair, A-T and its reverse T-A, G-C and its reverse C-G, a typical gene being about a thousand base pairs long. The base-pairs were subsequently shown to be grouped in triplets with each triplet specifying a particular member of a set of 20 amino acids according to the so-called genetic code, the whole gene being a blue-print for the construction of a particular chain of amino acids, a protein or polypeptide. It is through the active chemical properties of its coded polypeptide that a gene expresses itself and is biologically significant.

A mutation to a gene could now be seen to consist in one or more base-pairs being changed to another member or members of the set of four possibilities, A-T, T-A, G-C, C-G, this happening to the initial cell at the germination or conception from which an individual of a species was derived. The chance of such a change occuring due to a copying error was measurable, and was found to be about 10^{x} per base-pair per generation–i.e. about 10^{5} for any base-pair to be changed for a whole gene with a thousand base-pairs. This result was a death knell for neo-Darwinians since it forced evolution according to their views to be a one-step-at-a-time affair, a requirement which both experiment and commonsense showed to be impossible.

Figure E2.1 is a schematic representation of the mode of operation of an enzyme. An enzyme is a polypeptide which coils into an approximately spherical shape but with a highly specific site at its surface, a site shaped to hold the chemical substances in the reaction which it catalyses, chemical substances existing in many cases outside the biological system itself, chemical substances which do not evolve with the system. This fitting to the shapes of externally-defined substances is a constraint an enzyme must meet in order that it should fulfil its biological function. Exactly how many of the hundred (or several hundred) amino acids in the polypeptide chain of an enzyme must be explicitly defined in order that this shape criterion be satisfied is a matter for debate, but the number cannot be trivially small. If it were so, there would surely be far more variability of structure in the enzymes found catalysing the same chemical reaction in bacteria, humans, and in a potato. The number of amino acids in an enzymic polypeptide chain that cannot be changed without destroying the function of an enzyme is probably at least a half (see entry E9) and may in some cases be considerably more than a half. This demands that hundreds of base-pairs be appropriately placed in the gene which codes for the enzyme. If one is given an initial situation in which these requisite base-pairs are already correctly placed, well and good, but if the requisite base-pairs are not correctly placed initially, it is essentially impossible that copying errors will ever lead to a functioning enzyme. The difficulty is that all the key base-pairs have to come right simultaneously, not one-at-a-time, because there is nothing to hold individual base-pairs right until the whole lot are right. Every $\sim 10^8$ generations the key base-pairs are randomly shuffled, with the consequence that as some come right others go wrong. The chance of n requisite base-pairs happening to come right at each random shuffling is 4^n, so that with $\sim 10^8$ generations required for a shuffling the number of generations needed for a mutational miracle leading to a functioning enzyme to occur is $\sim 10^8 \cdot 4^n$, which for n of the order of a hundred is a lot of generations. But not too many for the neo-Darwinians, who know their theory to be right by some kind of revelation, and who therefore are not embarrassed to offer the most unlikely proposals in its defence. How they continue to argue in the present situation is discussed further in entry O3.

E 3. Punctuated Equilibria or Punctuated Geology ?

If it were possible to circumvent the criticism of neo-Darwinism given at the end of entry E2, arriving at the complex structures of genes several hundreds of base-pairs long by mutations that obtained correct pairs one-at-a-time, with natural selection somehow holding each pair fixed as it came right, evolution would necessarily have to proceed in a very large number of tiny steps, hundreds of steps for each of tens of thousands of distinct genes. There would be two ways to support this point of view. If both worked out well, one would be obliged to respect the neo-Darwinian position, but both ways turn out badly, as the criticism given at the end of entry E2 warns that they inevitably will. One way would be to demonstrate the mathematical validity of a small-step genetical theory (discussed in entry E5) and the other would be to obtain   direct evidence from the paleontological record showing that markedly separated stages in an evolutionary chain are linked by many intermediate small steps. So far from this being found, new species arise abruptly in the paleontological record, forcing the neo-Darwinian theory again onto the defensive in exactly the place where it might hope to be strongest if it were true.

Defensively, it has been pointed out (for example, recently by T. H. **van Andel**, *Nature*, 294, 1981, 397) that present-day sedimentation rates, if maintained throughout geological history, would have resulted in greater depths of sediments than are in fact found from the various geological periods, implying it is argued either much erosion of sediments, in which case the fossil evidence has been largely destroyed, or it might have been that there was a cessation of sedimentation over much of geological history, in which case the fossil record would have been established only sporadically. Evolution in small steps could then be made to appear as a sequence of jumps, simply by the discrete manner in which the evolution happens to be recorded in the presently available fossil record.

All this might be possible as a defensive manoeuvre, but the argument lacks the force of proof. When a curve is drawn through a number of points, the points themselves need occupy only a small fraction of the total range of the **abcissa** – what matters for constructing a curve is that there be enough points and that they be suitably distributed **with respect to the form of the curve itself. Moreover, sediments are available from many** geographical areas, and gaps in one place can be filled by available sediments in another place, unless erosion or a lack of sedimentation invariably conspired to be contemporaneous over all areas. For **small** evolutionary changes such a complementary association of different areas might be considered difficult to achieve but if we are looking for big changes, as from reptiles to mammals for example, a geological resource of this kind should be possible. One could see the defensive argument working in particular cases, but it is implausible to require it to work in every case, as it would need to do to explain the general abruptness of emergence of new species.

If, on the other hand, evolution really does proceed in sudden steps which separate extended time intervals of near-constancy, punctuated equilibria as such an evolutionary process has been called, one would expect to find examples of abrupt changes within continuous ranges of sediments. The question of whether sediments were really laid down continuously or discretely in the manner discussed above, is a matter for the judgements of professional geologists and palaeontologists. If we have understood their findings correctly, punctuated equilibria exist (for example, P. G. Williamson, *Nature*, 293, 1981, 437).

Although neo-Darwinians appear to have convinced themselves that they can explain such findings, we are at a loss to understand their point of view. One might attempt to conceive of many small mutations being accumulated during a time interval of near-constancy of a species, of the mutations establishing a potential for sudden change in a species, like the slow winding of a catapult and of the catapult eventually being suddenly released. But many small mutations established without regard for selective control would mostly be bad, and if there was indeed selective control we should simply be back again with the previous state of affairs, slow evolution in small steps, not punctuated equilibria.

Large advantageous mutations could explain the findings, but large advantageous mutations requiring many base-pair changes in the DNA structure of a gene or genes are exceedingly improbable for the reasons discussed at the end of entry E2. Large advantageous mutations requiring only a few base-pair changes might be postulated, but this would be to suppose that genes hover on the edge of marked advantage for species without natural selection having established them in such a critical position. In effect, a *deus ex machina* would be implied. In effect, the theory would have become open in the sense of entry E2, not closed as it is supposed to be in the neo-Darwinian theory. The position then comes close to our own point of view, to be explained in entry E4.

Could abrupt changes to a species be caused by sudden geological changes one might ask ? Only to the extent that changes in the physical environment produced selection with respect to the already-existing varieties of species. We should then be back with Patrick Matthew in 1831 and Edward Blyth in 1835 (entry E1). Geological changes could release genetic potential in the sense explained in entry E4, but geology cannot create genetic potential.

E 4.   Evolution by Gene-Addition

The concept of higher and lower animals, higher and lower plants, is widespread throughout classical biology, and it can be given objective definition in terms of greater or lesser degrees of complexity in the **organisation** and function of living forms. It is safe to say that if the biologists of the first half of the present century had been asked to guess the relative quantities of genetic material present in various forms general opinion would have favoured a strong positive correlation between quantity and complexity of function, the higher the plant or animal the greater the amount of genetic material. Figure E4.1 shows the results of actual measurements, the one part of the figure for animals, the other for plants, with the various taxa ordered generally with respect to complexity of function (A. H. Sparrow, H. J. Price and A. G. Underbrink, in *Brookhaven Symp. Biol.*, 23, 451, 1972). Except that procaryotes do have significantly fewer base-pairs than eucaryotes, and viruses have still less than procaryotes, the expectation is not borne out. The lungfish easily outclasses the human in the number of its base-pairs. Who would have guessed that the amoeba *chaos chaos* **would** have had five hundred times more genetic material than the primates ?

It might seem odd that the ideas on evolution held by neo-Darwinians have managed to survive Figure E4.1. One might have expected this remarkable new data to have sparked at least one or two revolutionary ideas. The reason for this congealed state of affairs is simply that the usual evolutionary theory explains little or nothing anyway, so that a further mysterious set of facts scarcely makes an already unsatisfactory theory much worse. It is only good theories that can be upset by new facts. A dead horse can take any amount of beating.

Fig. E. 4.1. DNA content per cell and per chromosome of various organisms.

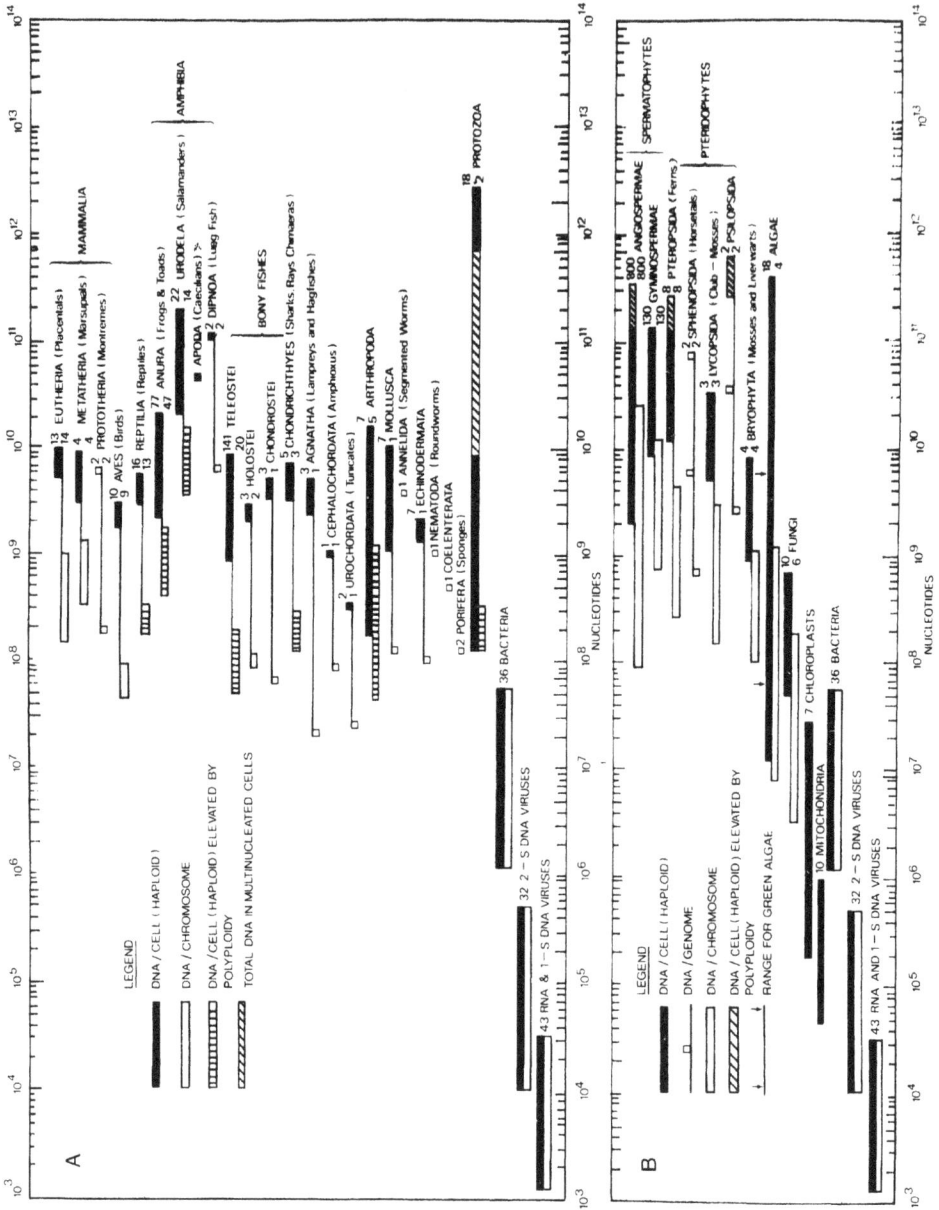

Evidence that microorganisms are continuously incident from space was considered in entries D1 to D8, it being argued that such microorganisms are most readily detected through a component which is pathogenic to terrestrial organisms. Viruses and viroids were considered as well as bacteria, microfungi and protozoa. Some commentators (not professional virologists, at least not to our faces) have claimed that pathogenic viruses cannot be incident from space, for an imagined reason which they believe overrides the many facts which prove otherwise. The argument seems on minimal thought to have the attractive quality of a one-line disproof. Viruses are specific to the cells they attack it is said, as if to claim that human viruses are specific to human cells. While a minority of human viruses might be said to be specific to the cells of primates, most human viruses can actually be replicated in tissue cell cultures taken from a wide spectrum of animals, some indeed outside the mammals entirely. The proper statement therefore is that viruses are generally specific to the cells they attack to within about 150 million years of evolutionary history. Actual diseases tend to be specific to particular species it is true, but this is not the same question, which appears to be where confusion has arisen in the minds of some critics. The ability of a virus to produce a clinical attack of disease in a multicellular plant or animals involves the special physical structure* and the particular immunity system of the creature under attack, and possibly other factors also,** all of which are irrelevant to whether the virus can attack individual cells.

If we had knowledge that evolution was an entirely terrestrial affair then of course it would be hard to see how viruses from outside the Earth could interact in an intimate way with terrestrially-evolved cells, but we have no such knowledge, and in the absence of knowledge all one can say is that viruses and evolution must go together. If viruses are incident from space then evolution must also be driven from space. How can this happen ? Viruses do not always attack the cells they enter. Instead of taking over the genetic apparatus of the cell in order to replicate themselves, a viral particle may add itself placidly to one or other of the chromosomes. If this should happen for the sex cells of a species, mating between similarly infected individuals leads to a new genotype in their offspring, since the genes derived from the virus are copied together with the other genes whenever there is cell division during the growth of the offspring. Viroids, consisting of naked DNA and perhaps representing only a single gene, penetrate easily into cells, and their augmentation of the genotype may well be still more important than the addition of viruses.

* The herpes virus can attack brain cells. Fortunately, this does not happen normally because the virus is not permitted physical access to the brain, otherwise its effects would be widely lethal.

** It seems possible that attacks of disease are in some cases triggered by a space-borne viroid rather than by the fully-fledged virus. The fully-fledged virus is the *output* from diseased cells, and it is conceivable that the output from cells contains genes derived from the cell itself. The output would then be more specific to the cell than was the original trigger. There are indications that the special peculiarities of influenza may be due to this kind of process (in *Space Travellers*, 1981, University College, Cardiff Press, page 171). The discussions of entries D3 and D4 would not be affected if this were so.

Genes newly obtained in this way may have no evolutionary significance for the plant or animal which acquires them, and for the majority of new genes this would quite likely be so, because each life-form will tend to pick-up a random sample of whatever happens to be incident upon it and in the main a gene acquired at random will probably find no useful genetic niche. It will simply replicate with the cells of the life-form in question without yielding a protein of relevance to the environmental adaptation of the species ; indeed, if the gene remains unaddressed in the operation of the cell, it will not yield any protein at all. It will remain ' unexpressed ' as one says. So we deduce that many of the genes present in the DNA of every plant and animal will be redundant, a deduction that is overwhelmingly true. Some 95% of the human DNA is redundant. Even higher percentages are redundant in lower animals, which goes some way towards an understanding of how it comes about that a lowly creature may nevertheless have an enormous amount of DNA (Figure E4. 1).

A gene that happens to be useful to the adaptation of one life-form may be useless to another. Incidence from space knows nothing of such a difference, however, the gene being as likely to be added to the one form as the other. So genes that become functional in some species may exist only as nonsense genes in other species. This again is true. Genes that are useful to some species are found as redundant genes in other species. Suppose a new gene or genes to become added to the genotype (genome) of a number of members of some species. Suppose also that one or more of the genes could yield a protein or proteins that would be helpful to the adaptation of the species. The cells of those members of the species possessing the favourable new genes operate, however, in accordance with the previously existing genes, and since the previous mode of operation did not take account of the new genes, a problem remains as to how the new genes are to be switched into operation so as to become helpful to the species. This question is discussed in entry E6. Here we simply note that, because there is no immediate process for taking advantage of potentially-favourable new genes, such genes tend to accumulate unexpressed. As potentially-favourable genes pile up more and more, a species acquires a growing potential for large advantageous change, it acquires the potential for a major evolutionary leap, thereby punctuating its otherwise continuing state of little change—its ' equilibrium ' (entry E3). This is why new species appear abruptly, a concept that will be developed further in entry E6.

E 5.   Genetics in Open and Closed Systems

According to our point of view essentially all genetic information is of cosmic origin. The information does not have to be found by trial and error here on the Earth, so that mutations in the sense of the **base-pair** shufflings discussed in entry E2 do not have the positive relevance for us that they have in the neo-Darwinian theory. Indeed, just the reverse. Base-pair shufflings are disadvantageous because they tend to destroy cosmic genetic information rather than to improve it, and this is especially so during the interim period before advantageous new genes are switched into the ' program ' of a species (entry E4), before they become protected from serious deterioration by natural selection. In neo-Darwinism on the other hand, systems are closed, they start with no information and seek somehow to find it, whereas open systems start with high-grade genetic information which it is important for them not to lose.

For this latter point of view the base-pair copying-error rate should be as low as possible, while for the neo-Darwinians it needs to be high if the requisite sophisticated information is ever to be found, just as the monkeys with their typewriters need to work exceedingly fast if they are to arrive within even a cosmic time-scale at the plays of Shakespeare. The copying-error rate is in fact very low, DNA is very stable, clearly supporting the position of entry E4, not that of the neo-Darwinians.

Since many people think neo-Darwinism to be established beyond doubt, and the questioning of it an act of sacrilege, it is worth leading that theory to the knacker's yard yet again, which will be done in the present entry. We shall now show that even within its own postulates neo-Darwinism is self-contradictory. At the end of entry E2 the neo-Darwinian theory was shown to require each important base-pair of every gene (initially not correct) to be held by natural selection when it eventually becomes miscopied to the correct form. The ' discovery ' of genes has to be a one-step-at-a-time process, otherwise there is no possibility worth speaking about of all the many base-pairs coming to their required forms simultaneously. If neo-Darwinism is to be consistent with the detailed structures of genes it is therefore essential that evolution proceeds in very many small steps.

This need to proceed in small steps was already guessed by mathematical geneticists in the first quarter of the present century (e.g. R. A. Fisher, *The Genetical Theory of Natural Selection*, Oxford, 1930). Looking back at this old work it is surprising to find advantageous results for the neo-Darwinian theory being claimed, when even quite easy mathematics shows otherwise, especially as the claimed results were an affront to commonsense. When a mutation is small, its effect on the performance of an individual is so marginal that it scarcely affects the number of offspring born to the individual. Is natural selection really so powerful that in such marginal situations it can stamp-out the flood of slightly negative mutations while preserving the trickle of slightly positive ones ? Commonsense says no, and commonsense is correct, as we shall shortly demonstrate.

The remedy of R. A. Fisher was to postulate that small negative mutations are not more frequent than small positive ones, but this supposition also defies commonsense, because it is a matter of experience that complex organisations are much more likely to develop faults than they are to find improvements, a view well-supported by modern microbiology. If the identities of only a hundred base-pairs per gene are important for an animal with 100,000 genes, there are ten million ways at each copying of going wrong.

75

With an error probability of $\sim 10^{-8}$ per copying per base-pair, the chance Q of a significant deleterious mutation occurring per generation per individual is $Q \simeq 10^{-1}$. For a breeding group with N members, the number of deleterious mutations injected into each generation is 2QN, which for a typical breeding group, say N = 10,000, gives two thousand deleterious mutations per generation, quite a burden to be carried every few years. The number of advantageous mutations must surely be much less than this.

An example will make the situation clearer. Suppose a printer sets up a page of 400 words with a dozen spelling mistakes among them. A single letter somewhere on the page is changed at random, thereby introducing a small ' mutation '. The chance that such a mutation will make the spelling worse, giving thirteen mistakes, is evidently overwhelmingly greater than that the mutation will just happen to correct one of the initial dozen errors. Except that genetically there are only four letters for a base-pair (A–T, T–A, G–C, C–G) instead of the twenty-six letters of the English alphabet, the cases are not unfairly compared, especially as the greater number of letters in the literary case is more than offset by the far greater number of genetic 'words', 100,000 genes, any one of which can go wrong.

Since we have analysed the mathematical problem elsewhere (*Why Neo-Darwinism Doesn't Work*, University College Cardiff Press, 1982) it will be sufficient to quote the main results here. In the case of an individual with an advantageous dominant mutation **present on either set of chromosomes write 1 + x for the ratio of the average number of offspring produced** to the average number of offspring for others without the mutation. Then the fraction of such mutations which natural selection spreads through the entire species is about 2x. Thus for x = 0.001, a fairly considerable advantage of 0.1 percent, the chance of a mutation spreading through the species is no more than 1 in 500. It therefore needs some five hundred fairly considerable mutations, each of them likely to be a rare event, before just one is retained by the species. Hence for mutations with x small, natural selection adds up very little that is good.

The trouble lies in stochastics, an effect that was inadequately considered by the **early mathematical geneticists. For a heterozygote with respect to a gene of small x there is already nearly a 25 percent chance that the mutation in question will be lost in the first generation, simply from the random way in which the heterozygote allots one or other of its duplicate set of genes to each of its offspring. In the second generation there is again a chance of about 3/16 that the mutation is lost. Stochastics consists in adding up and allowing for these extinction possibilities, which greatly dominate the effects of natural selection when small mutations first arise.**

For the same reason natural selection by no means removes all that is bad, as classical biologists supposed. For deleterious mutations it is the recessive case that matters most. If for simplicity of argument one takes all recessive deleterious mutations to be equally bad* an elegant result can be proved. Subject to the disadvantage factor x being sufficiently small, the rate at which deleterious mutations spread through a whole species is equal to the rate Q of the mutations per individual**, just the same result as was proved about a decade ago for neutral mutations (M. Kimura and T. Ohta, *Genetics*, 61, 1969, 763).

If natural selection fails for moderate mutations to add-up more than a small fraction of what is good, and if natural selection fails to exclude a damaging fraction of the much more frequent disadvantageous mutations, how can species ever become better adapted to their environment ? For small-step mutations they cannot, which is why neo-Darwinism fails genetically, why positively-evolving systems must be in receipt of genetic information from outside themselves, as was discussed in entry E4. The best a **closed system** can do is to minimise in *disadaption* to the **environment**, a topic that is discussed in entry E7.

Natural selection works excellently for open systems, since with high-grade genetic information coming from outside a system, advantageous changes have large values of x, with 2x of order unity, so that if such a change occurs for only one or two individuals of a species, natural selection operates to fix the change throughout the entire species. Such major advantageous steps have to occur with a sufficient frequency to more than offset the numerous small deleterious mutations which still produce disadaptation at the rate discussed above. In   effect, the situation is a race between uphill jumps produced by externally incident genetic information and the downhill slide of the already-existing genes, which natural selection can only moderate but not remove **entirely**. This produces a highly fluid situation, with species either advancing rapidly or sliding backward towards extinction-as is observed to have happened for the higher plants and animals.

When one looks back at the mathematical geneticists of the first half of the present century, it is clear they approached their work in the complete conviction that the neo-Darwinian theory was correct. As the majority of them saw it, their duty was to explain why a theory known to be correct was indeed correct, a mode of argument not unlike a chemist attempting to work backwards through an irreversible reaction, or like an inept student in an examination trying to work backwards from the answer to a problem to its mode of solution. This wrong-headed approach led somewhat naturally to a

---

*For a deleterious mutation write 1-x for the ratio of the average number of offspring produced by an individual with the mutation on both chromosome set 0 to the average number of offspring produced by individuals without the mutation. The disadvantage factor x (> O) is taken the same for all deleterious mutations.

**The condition on x is that the product of x and the number N of individuals which constitute a randomly-mating breeding group be not greater than ˜ 1. This leads to a *disadaptation* factor exp (–QG/N) arising in G generations. For values of N appropriate to mammalian species this disadaptation factor becomes an embarrassment to neo-Darwinian theory as G increases above a million generations.

prostitution of logic which was mercifully concealed from the public in a haze of mathematical symbols. The irony is that the correct answer was easy to find if only the mathemetical geneticists had troubled to look for it in the right direction.

E 6.   Favourable Mutations in Open Systems

Open systems do not have to find genetic information *de novo*, because they are in receipt of genes from outside themselves. However, newly-acquired genes must lie fallow for a while, since the mode of operation of the cells of the species in question cannot 'know' in advance of their arrival. The sequence of events whereby genes are used may usefully be described as the cell program. What needs to be done therefore to promote evolution in an open system is to alter the cell program to take into its operation new genes which it did not use before. The problem to be considered here is the logic of this situation.

A cell program may be thought of as analogous to a computer program. With computers, the program is something different from data and from the closed subroutines which constitute the backing storage. Computers can be operated on many different programs using the same physical hardware and the same backing facilities – examples of the latter are routines for taking logarithms and integrating differential equations. Something of the same kind almost surely exists in biological systems. Genes for the production of enzymes, haemoglobin, the cytochromes, are examples of subroutines that run across all of biology. It is even the case that genes capable of producing some of these standard products, haemoglobin for instance, exist in life-forms which normally make no use of them, just as standard computer languages like FORTRAN or BASIC contain more facilities than are used in any particular individual program.

In days long ago, before sophisticated computer languages were available, when it was necessary to remain closer to the electronic nature of the computer itself, one was perhaps more keenly aware of the distinction between the logical instructions which constitute a program and the numbers or words on which the program operates, even though both were stored in the computer in exactly the same way, as sequences of bits. Although numbers and logical instructions were similar electronically, you could not use numbers for logical purposes or process your logical instructions arithmetically (a few very slick fellows tried and were sometimes successful, but the tricks of this particular trade were too subtle to have survived into current practice). As well as numbers constituting data and logical instructions making up the program, something else was needed, a starting point and an end point, birth and death.

Do biological systems operate in a similar way ? Are the logical instructions constituting the cell program stored as genes, but used quite differently from the genes which code for working polypeptides such as the enzymes ? Is everything stored as base-pairs in the DNA, just as everything in a computer is stored in sequences of electronic bits ? It is tempting to suppose so, but there are indications that it may not be so. The DNA of a chimpanzee is extremely similar to that of a human. Therefore the scope for producing working polypeptides is essentially the same in the chimpanzee as it is in ourselves. Thus the chimpanzee and the human look like two different programs operating on the same physical hardware, on the same backing storage as one might say. If the different programs were on the DNA we might expect to see less close similarity, less homology, between the base-pair sequencing of the two species, unless program storage occupies very little of the DNA, unless the logical ordering which makes us specifically human and a chimpanzee specifically chimp is in each case rather trite and short. Perhaps the logic of being human is rather trivial, but one **prefers** not to think so.

A less subjective objection is that DNA seems far too stable to be the source of the cell program. If the cell program were so contained, body cells could be replicated a very large number of times without the program being much impaired, permitting animals to have exceedingly long lives, whereas the evidence shows that the program becomes seriously muddled after only a handful of replications. Recognizing this discrepancy some biologists have argued that senesence is itself a deliberate part of the program, deliberate in the sense that natural selection has prevented us from living long by explicitly stopping the coding of essential working polypeptides. This opinion is to be doubted, however, because wild animals commonly die violent deaths before their time is run, so there is no cause in nature for natural selection to prevent lives from being too long. Yet all animals do show senesence, if artificially protected against violent death most of them even more markedly than we do, indicating that senesence is not artificially contrived. The implication is that storage of the cell program must be ephemeral. It is preserved with reasonable fidelity in gametes, but soon runs down and becomes forgotten, leading to grey hair and the like, as soon as the somatic cells are required to replicate more than about a hundred times.

If a person tells you that the telephone number of a mutual acquaintance is 814973 and you immediately commit the number to paper you have it in stable storage, like base-pairs on DNA. But if you seek to remember the number aurally in your head, it will be gone at the first distraction, a knock on the door or a pan of milk boiling over on the stove. This seems to be the way of it with our cell program. Once we have lost it, the thing never comes back, although if it really is retained in our gametes somebody may succeed someday in copying it back into our somatic cells, with interesting sociological consequences.

In spite of these difficulties, suppose for a moment that those who think the cell program is written on the DNA are correct. How would the program actually do something ? Not by merely remaining on the DNA, because DNA by itself is inert. The program would need to be translated into polypeptides and it would be the polypeptides that really did something. So why not let the program be polypeptides in the first place ?

Or if not the whole program, suppose an essential part of it is in polypeptide form, without there being any reference genes on the DNA form from which the initial polypeptides can be recopied if they become lost. One might conceive for instance that the initial polypeptides comprise a catalogue of what in computer terminology would be referred to as calling sequences, which is to say some means of determining so-called introns for finding important genes on the DNA. Senescence looks very much like the progressive garbling of the entries in such a catalogue, so that we end in old-age by not being able to find more than a small fraction of the genes necessary for vigorous life. All this is relevant to the evolutionary problem set out at the beginning, since the less rigidly fixed the cell program the more readily one can conceive of it being changed. The change needed for an evolutionary step must involve some means of addressing new genes added to the DNA, the genes which supply the potential for an evolutionary leap. This means actually doing something, not just adding DNA blueprints for doing something at some stage in the future. Actually doing something means polypeptides, and doing something new means new polypeptides, which implies a working addendum to the old cell program. Where one now asks is such a working addendum to come from ? Only it seems from a virus.

When a virus invades a cell it mostly happens that the virus multiplies itself at the expense of the invaded cell, which it does by stopping the old cell program and inserting its own program, both necessary but not sufficient properties for what we are seeking. The several viral particles thus produced then emerge from their host in search of still more cells to invade, and so on apparently *ad infinitum*. This behaviour is usually viewed as a permissible oddity of biology, permissible because the virus survives, and survival is all according to the opinions of neo-Darwinians. Yet mere survival leaves the virus as a disconnected organism without logical relationship to anything else. Once one admits, however, that logical relationship is at least as valid a concept as survival, indeed that survival is impossible for any organism without logical relationship, the situation becomes different. The virus becomes a program insertion with the essential capability of forcing cells to take notice. Many such program insertions are needed to cope with many stages of evolution for many creatures, both on the Earth and elsewhere. Hence many viruses are needed, and even if the entry of a particular virus into cells is restricted to situations in which the cell program and the viral program match together in a general way, it will not usually happen that a virus on entering a cell has precisely the appropriate program insertion to suit the life-form in question exactly at its current stage of evolution. There will have to be many trials before precisely the current program insertion is found. So what is the virus to do in the majority of cases where the situation is not quite right ? Give up the ghost and expire ? If it did so, what about the other creatures somewhere in the Universe that may be in dire need of its particular evolutionary contribution ?

Viruses seek cells, not *vice versa*. Speaking anthropomorphically, they have the job of driving evolution. They cannot give up the ghost and expire, otherwise nothing would happen, the situation would be as dead as mutton. So they augment themselves by increasing their number and then they press on, forever seeking to find the cells where they are needed. As soon as one looks for logical design, the situation immediately makes sense. Besides which, the infective ability of viruses also plays a crucial logical role. For species with a sexual mode of propagation there is a big question mark as to how an evolutionary leap could ever be possible, because the same leap must occur in at least one male and one female, otherwise the male and female gametes will not match properly, and there will be reproductive trouble in the second generation, if not indeed immediately. Since the probability of an evolutionary leap occurring is small, requiring first a building of a potential for the leap and then finding the correct addendum for the cell program, it would be a poor result if the individual for whom all this happened were then to be sterile. Yet if we need the same improbable sequence for the opposite sex also, the small probability is squared, and moreover the changed male living in London would then have the problem of finding the changed female living in New York, making such an uncorrelated situation quite hopeless. The solution to this last problem is infectivity. The same changes, all being virus induced, can be infective between individuals in close contact at the same geographical location, and in this case the small probability is not squared, and moreover the similarly affected individuals are automatically together and so cannot avoid finding each other. The logic of an evolutionary leap demands infectivity. Infectivity also explains why after an evolutionary leap the previous line does not persist, since with an evolutionary improvement sweeping through a species like a disease, a negative disease as one might say, the previous line is overwhelmed by the superior adaptation to the environment of the drastically changed creatures. Only in this dramatic way can evolution counter the degenerative effect of the small but steadily-occurring miscopying of genes, the downward drag that was mentioned above and is considered in more detail in entry E7.

The above discussion also makes it clear why viruses have to be generally specific to the cells they invade (entry E4).

E 7.   The Survival and the Extinction of Closed Systems

Here we accept the conclusion of entry E5, that natural selection is not able to fix in a species more than a small fraction of the infrequent advantageous mutations which arise through the shufflings of base-pairs on the DNA, and hence that internal processes cannot improve the adaptation of a species sufficiently to be significant. Only by importing genetic information from without can adaptation be improved in an important degree, and this we consider in the present entry to be absent.

Although natural selection (together with stochastic processes) remove a large fraction of the numerous deleterious mutations, sufficient of them necessarily remain to degrade the adaptation of a species quite seriously. The most troublesome deleterious mutations are the recessives, which arise because initially useful polypeptides change gradually into nonsense proteins as random shufflings of the base-pairs alter their amino-acid sequences to less useful arrangements. A deleterious recessive on the same gene of both chromosome sets of a diploid cell has a disadvantage expressed by the average of the ratio of the number of offspring produced by such individuals to the number of offspring produced by individuals without the mutation (but who are otherwise similar). Write this disadvantage factor as $1 - x$, so that x is a positive number between zero and unity.

The extent to which the combination of stochastic effects (entry E5) and natural selection permits such a mutation to penetrate a species depends on $4 xN$, where N is the number of diploid individuals making up the breeding group, taken to mate within itself at random. Write Q_1 for the average rate of occurrence of deleterious recessive mutations with $4xN > 1$. per individual per generation, and Q_2 for the average rate per individual per generation for mutations with $4xN < 1$. Starting from a pure line state of affairs in which all chromosome sets are identical throughout a species, the situation which transpires is the following. Over a very long time-scale the mutation rate Q_2 degrades the quality of the pure line while on a shorter time-scale the rate Q_1 degrades the species relative to the slowly changing pure line by the factor $\exp{-Q_1}$. We discuss these two distinct effects separately, after noting that both Q_1 and Q_2 are generally of order unity. Taking mammals as an example, each diploid has $\sim 6.10^9$ base-pairs, so that with a copying error rate of $\sim 10^8$ per base-pair per generation there are 60 miscopyings per individual per generation. However, only those miscopyings of pairs belonging to expressed genes are relevant in the present connection, say 5 per cent of the total, giving ~ 3 relevant miscopyings per generation, i.e. $Q_1 + Q_2 \simeq 3$, taking most of the miscopyings to be deleterious and most of them to be of a recessive nature. In the absence of information as to how these ~ 3 miscopyings should be divided between Q_1 and Q_2 we assign them equally, $Q_1 \simeq 1.5$, $Q_2 \simeq 1.5$, both per individual per generation.

Suppose for the moment that all mutations contributing to Q_1 have the same value of x. Stochastic effects give a chance $\sim \sqrt{N/x}$ of each such mutation spreading into $\sim \sqrt{x/N}$ members of the species. Thereafter natural selection operates to prevent further spreading. Indeed natural selection works to reduce the number of distinct mutations which become spread by stochastics, while the injection of new mutations works to increase the spreading of distinct mutations. An equilibrium between these opposing effects becomes established in $\sim 2\sqrt{N/x}$ generations, an equilibrium in which $\sim 2 Q_1 N$ distinct mutations are each spread at random in $\sim \sqrt{N/x}$ members of the species, giving an average of $\sim 2Q_1 \sqrt{N/x}$ mutations per diploid.

Because of the randomness with which the distinct kinds of mutations are distributed, the mutations on the two chromosome sets of a diploid cell are uncorrelated, so that the same gene is affected on both chromosome sets only by **chance**, the chance of a coincidence being $\sim 1/2 \sqrt{Nx}$ for each of the $\sim Q_1\sqrt{N/x}$ kinds of mutation that on the average are present on every chromosome set. For a diploid there are thus $Q_1/2x$ deleterious recessive coincidences. Each of the N individuals forming the breeding group therefore encounters a reproductive penalty relative to the initial pure line expressed by the factor $(1-x)^{Q_1/2x}$ which for x appreciably less than unity in general is $\exp{-Q_1/2}$. Hence at a typical mating of a male and female, each with the degradation $\exp{-Q_1/2}$, the combined penalty is $\exp{-Q_1}$, as already stated above.

The value of x does not affect the penalty, only the rate Q_1 is relevant at which mutations arise per individual per generation. This remarkable result permits the assumption that all mutations contributing to Q_1 have the same x to be dropped. (If one had mutations with either x' or x'', $x' \prec x''$, the greater deleterious effect of an x' mutation would be compensated by the greater number of x'' mutations that penetrated the species.) Thus the penalty per mating pair relative to the original pure line is $\exp{-Q_1}$, with Q_1 now interpreted as the total rate of occurrence of mutations with $4xN > 1$ per individual per generation, a result which leads to the deduction that no closed species can have appreciably more than 10^8 expressed base-pairs on its DNA. Otherwise with a miscopying rate of $\sim 10^{-8}$ per base-pair per generation we should have Q_1 much larger than unity and the penalty $\exp{-Q_1}$ would be exceedingly severe, likely enough leading to an extinction of the species.*

Although all base-pairs are subject to much the same miscopying rate, only a fraction of the mutations which occur ever penetrate a species significantly. The majority of mutations are removed by stochastic effects. There is no means of determining which mutations happen to penetrate and which are eliminated almost immediately, the issue is a matter of chance. Thus if we imagine an initially pure line separated into two breeding groups, after a suitable number of generations have elapsed both groups will be afflicted by the **same** degradation factor $\exp{-Q_1}$. However, the recessive mutations causing this same degradation factor will mostly be different from one group to another.

Suppose in such a situation that the two groups are artificially mated together, as for instance two varieties of wheat might be crossed by a plant breeder. The factor $\exp{-Q_1}$, afflicting both groups separately, evidently disappears almost entirely from the first generation of hybrids, because genes affected by recessive mutations on the chromosome set derived from the one group do not in general match the mutations on the chromosome set from the other group. In other words, the mistakes of the one are

---

* This limitation on the number of expressed base-pairs assumes no gene duplication. The number could be increased by multiple polyploidy, for example.

shielded by the other, and with the degradation factor $\exp-Q_1$ thus disappearing from the hybrids the vitality of the original pure line is restored. However, coincidences of recessive gene mutations begin to appear again already in the second mixed generation, and random matings with chromosome cross-overs occurring degrades the situation in only a few generations about half-way back to what it was before. This is the phenomenon of hybrid-vigour well-known to plant breeders.

Of the total of NQ_2 mutations with $4 \times N \leqslant 1$ that arises in each generation, a fraction $\sim 1/2N$ penetrates a species entirely due to stochastic effects, thereby slowly changing the original pure line that provided the standard relative to which the degeneration factor $\exp-Q_1$ was measured in the above discussion. Hence the standard of reference itself deteriorates relative to the original pure line by a factor

$$(1 - \bar{x})^{Q_2/2} \quad (1 - \bar{x})^{Q_2/2} \simeq \exp(-\bar{x} Q_2) \qquad\qquad \text{E7.1}$$

per generation for each mating pair, $\bar{x}$ being the mean of x ($x \leqslant 1/4N$). This further source of deterioration is cumulative from generation to generation ; after G generations it becoms $\exp(-\bar{x} Q_2 G)$. Talking $Q_2 \simeq 1.5$ as indicated above, there is a decline by e^{-1} in $2/3\bar{x}$ generations, which for $\bar{x}$, say, equal to $1/6N$ is $\sim 4N$ generations. A number of interesting conclusions can be drawn from this result.

Under the condition assumed in the present entry, namely zero input of genetic information from outside itself, a species with N no larger than 10^5 is exposed to an overwhelming threat of extinction. Thus in a geological period of $\sim 10^8$ years with G upwards of 10^7, $\exp(-G/4N) = \exp-25$, surely a disastrous decline. Curiously for closed species with breeding groups no large than 10^5, it would be better if there were no small mutations, better if all deleterious mutations had $4 \times N \not> 1$, because natural selection could then prevent mutations from becoming fixed, and so could prevent the reference standard from deteriorating. The maximum penalty from deleterious recessives would then be $\exp-Q_1$, which is not cumulative from generation to generation. Continuing, however, with $Q_2 \simeq 1.5$, for a species to survive over a geological time-scale one at least of the following conditions must be satisfied :

(i) The breeding group N is very large, say $\sim 10^8$ or more.

(ii) The species is open to the receipt of genetic information from outside itself, and this external impulse is sufficient to upgrade the species at least as fast as it is being downgraded by internal mistakes.

For the larger mammals in the wild (i) is not satisfied, so that (ii) is necessary for long-term survival, as well as for the evolutionary development of mammals (entry E6). If plants and invertebrates are considered to be closed systems, then (i) must be satisfied.

Any closed species for which N falls appreciably below 10^x is doomed to extinction on a geological time-scale, and this no doubt is the reason why so many species have in fact become extinct throughout the geological record.

E 8. The Origin of Closed Systems

Closed systems degrade with time towards extinction, instead of showing the evolutionary adapatability of open systems. From where, one naturally asks, did closed systems derive their organised states in the first place ? From being open systems at some earlier time one naturally answers, raising the further question of whether the open condition occurred on the Earth or somewhere else in the cosmos.

Species are not in control of their own destiny. Control comes from viruses incident from space (entry E6). The arrival of viruses matching the cells of a species gives an open condition, while the absence of suitably specific viruses requires a species to be closed. One can suppose the incidence of viruses to have long-term variability, with the viruses in one geological period causing rapid evolution in a particular group of species, the mammals at present for instance, but leaving another group of species closed and so drifting slowly towards extinction. In another geological period the situation could be different, perhaps with the viruses related to species in a reversed way. Hence the evolutionary record might be expected to show great diversity, which it does, with one taxonomic category surging ahead, another lagging, in a constantly changing pattern.

This picture is so simple and satisfactory that it might be thought unnecessary to consider the more complex possibility of multicelled species incident as a whole from space, species which behave once on the Earth as closed systems. Yet the very invertebrates we suspect to be closed turn out to possess two properties essential for invaders from space, properties which seem unnecessary for terrestrially evolved species.

There is a limiting-size criterion for organisms from space, arising from the requirement that flash-heating on entering the terrestrial atmosphere must not be so severe as to destroy viability (entry A1). This criterion admits viroids, viruses, bacteria, colonies of bacteria and eucaryotic cells but it certainly excludes eggs with large embryos like those of a bird. Since creatures of comparative complexity, like birds, require embryos of considerable size, no way of getting such creatures on to a planet from outside seems at first sight to be possible. Yet there is a way. Send an egg containing the embryo of a simple creature accompained by the DNA of a more complex creature, and let the simple creature (after making safe landing) serve to accumulate the nutrients for the growth of the complex creature. In other words, let your creature have a larval state, as the invertebrates do.

At first sight also, it seems impossible to land from outside a species with bisexual propagation. The safe-landing of an egg at the upper limit of size, and with the need for the successful intervention of a larval state, is necessarily an event of low probability. If two such events are needed, one for the female, one for the male, the low probability becomes squared. The resulting far smaller probability must then be multiplied by a further small number, representing the chance of a male after safe landing ever finding a female after safe landing. So the end result appears to be a probability of the third order of smallness, much too minute to be feasible. Yet again there is a solution to the difficulty. Let either the male or the female have the ability to reproduce itself monosexually, parthenogenically, as invertebrates are again able to do. Then after a safe landing the single sex can build up so many copies of itself that the chance of the opposite sex finding any one of the copies, increased proportionately to the number of copies, is no longer unacceptably small.

The strange properties of parthenogenesis and of the existence of the larval state, which appear as oddities—freak properties—when viewed terrestrially, becomes essential if multicellular bisexual organisms are to populate the Earth successfully from space. The fact that such properties exist, and in the very organisms which seems closed in the sense of entry E7, gives one cause to ponder, if not indeed to become convinced.

E 9. Phylogenetic Trees

Sugars are broken down in all cells by a process of respiration involving about fifty enzymes. Of these fifty about twenty working together in a chain, known as the cytochromes, serve to move hydrogen atoms along the chain. The third member of the chain, cytochrome–c, is a polypeptide with 104 amino acids. Because it is a relatively small polypeptide and is comparatively easily isolated, the detailed ordering of amino acids in cytochrome–c has been studied with particular care (*Atlas of Protein Sequence and Structure*, Vol. 5, National Biomedical Research Foundation, Washington, D. C., 1972). Although there are variations of the amino acids in different organisms, 35 sites along the polypeptides were always occupied by the same amino acides. While this is less than a half of the total of 104 that was regarded in entry E2 as being fixed, the remaining 69 sites were by no means freely variable. Thus at 23 of them there could only be one of two particular amino acids and at 12 more sites only one of three amino acids, with some form of restriction at all but about 5 of the 69 sites. In total, this adds up to a greater measure of restriction than was considered in entry E2, where for simplicity the amino acids were treated as being either wholly fixed or wholly free.*

---

*If a half of the 104 amino acids were wholly fixed and a half wholly free, of the $(20)^{104}$ ways in which 20 different kinds of amino acids can be formed into chains $(20)^{52}$ would be biologically viable. The data indicate that about $(20)^{36}$ are viable equivalent to 68 sites being fixed, and 36 being wholly free.

Figure E9.1 was constructed by M. O. Dayhoff, C. M. Park and P. J. McLaughlin from this data for cytochrome–c. The hypothesis behind this figure is that all of terrestrial biology came from a common ancestor which had a postylated form of cytochrome–c represented by the tree trunk in the figure. At the first circle, encountered as one moves upward along the tree trunk, there was a supposed evolutionary divergence. Along the branch that went to the right, mutations caused changes in the ordering of the amino acids of cytochrome–c at 25x (104/100) sites along the polypeptide chain *, whereafter further branchings occurred, leading eventually at the outer tips of the branches to the cytochrome–c actually found in modern plants. Likewise for the branch to the left leading to fungi, yeast and neurospora. Along the central branch there were 15×(104/100) sites in the amino-acid chain at which changes took place before a branching occurred that led on the right to the cytochrome–c actually found in mammals, birds, reptiles and fish, and on the left to insects.

If one were given stocks of 20 differently coloured beads, strings, of the beads 104 sites long could be made up in $(20)^{104}$ ways. Choose one such string at random for each of the names printed in small type at the ends of the branches in Figure E9.1. Then examine the chosen strings with respect to the colourings of the beads, analogously to the identities at the 104 sites of the amino acids in cytochrome--c. From this artificial situation it would always be possible to construct a tree of the general form of Figure E9.1. Indeed it would be found possible to do so (for a particular choice of the set of strings) in very many different ways, and some criterion would be needed to single-out just one of the possible trees. An appealing criterion might be to choose the tree ˙or which the total of all the numbers appearing in it was least, the minimal tree as one might call it. Then one would have the analogue of Figure E9.1.

Since a minimal tree can be constructed artificially, how can Figure E9.1 tell us anything about biological evolution one naturally asks ? In itself, it cannot of course, but if we combine Figure E9.1 with additional knowledge the tree becomes relevant. From many other criteria we already know that fungi, mammals, insects, mammals, birds, reptiles and plants are markedly different from one taxonomic category to another. So what is interesting about Figure E9.1 is that a tree can be arranged so that species we thought similar are indeed close together in the tree, and taxonomic categories we thought widely different are far apart in the tree. Even so, this does not prove the evolutionary significance of the tree. It suggests evolution has occurred towards the tips of the branches, but the lower branches towards the trunk remain conjectural. So far as the data for cytochrome–c are concerned there need have been no ancestral cytochrome–c at all. The higher branches could have come from lines that were always separated here on the Earth, and perhaps even separated throughout the cosmos.

* The actual numbers of amino-acid changes have been multiplied by 100/104 to obtain the numbers of Figure E9.1. Hence one must multiply by 104/100 to restore the actually observed numbers. This form of normalisation is useful for comparing with trees obtained from other polypeptides having chains of different lengths.

Fig. E 9.1. Phylogenetic tree showing presumed evolutionary connection for cytochrome c. Each circle represents the sequence of a cytochrome c deduced to be ancestral to all species higher in the branches leading to that circle.

A minimal phylogenetic tree of this kind could, however, be constructed for every polypeptide of fixed length that occurs across many species. If there has been evolution from a common ancestor all such trees should have similar branches (although the **normalised** numbers on the branches would usually be variable). This expectation has recently been verified for mammals using five different polypeptides (D. Penny, L. R. Foulds and M. D. Hendy, *Nature,* 297, 1982, 197). But evolution among mammals was not in doubt of course. The doubt concerns the lower hypothetical branches of Figure E9.1.

Some steps towards obtaining information about the hypothetical lower branches have recently been taken, using the base-sequences of rhibosomal RNA rather than the amino acid sequences of polypeptides. The advantage of using 5 S RNA and 23 S RNA is that, unlike polypeptides such as cytochrome—c, the rhibosomal RNA is found in all microorganisms ; and since microorganisms have been regarded as the evolutionary precursors of the multicelled plants and animals, there would appear to be the opportunity (using a technique similar to the minimal tree) of coming closer to the supposed beginning of life. Results have not been encouraging for those who believe that all terrestrial life has emerged by evolution from an ancestral cell of terrestrial origin. No convincing common denominator between eucaryotes and procaryotes has been found. Even the procaryotes appear to have two distinct branches, the eubacteria and the so-called archaebacteria, between which no common trunk has yet been found (C. R. Woese, *Scientific American,* 244, 1981, 94 ; and G. E. Fox, E. Stackebrandt, R. B. Hespell, J. Gibson, J. Maniloff, T. A. Dyer, R. S. Wolfe, W. E. Balch, R. S. Tanner, L. J. Magrum, L. B. Zablen, R. Blakemore, R. Gupta, L. Bonen, B. J. Lewis, D. A. Stahl, K. R. Luersen, K.N. Chen and C. R. Woese, *Science,* 209, 457, 1980 ; also *Zbl. Bakt. Hyg., I* which contains papers given at the First International Workshop on archaebacteria, Munich, 27 June to 1 July 1981). While these latter results do not disprove the concept of a common ancestor, they make it well-nigh certain that, if there was a common ancestor, it did not originate on the Earth, a conclusion for which there is other evidence of a direct observational kind (entry M1).

I

Interstellar grains

I 1. Intersteller Extinction Data in the Visual and Ultraviolet

The interstellar extinction of starlight at wavelengths in the visual and ultraviolet have been collated by A. Sapar and I. Kuusik from a number of original sources (*Publ. Tartu Astrophysical Observatory*, 46, 1978, 71). For wavelengths λ longer than 2000 Å there is sufficient similarity between the original sources for the averages estimated by Sapar and Kuusik, marked in figure 11.1, to be sufficient for our purposes. For λ less than 2000 Å there is more variability however, and so two principal sources of data are also shown in figure 11.1, points in the figure marked (■) are from R. C. Bless and B. D. Savage, Astrophys. J., 171, 1972, 293, and those marked (▲) are from the *Ultraviolet Bright-Star Spectrophotometric Catalogue*, European space Agency, Paris, 1976.

A typical measured extinction value of 1.8 magnitudes at $\lambda = 0.55$ µm along a path length of 1 kiloparsec has been used to normalise the data points of figure I 1.1. Extinction of starlight may be due either to a scattering of the light from a star or to its absorption, with both the scattering and absorption being caused by the fogging effect of small solid grains of material lying along the path from the star to the earth. If I (λ) is the measured intensity in a small waveband centred at λ , and I_0 (λ) is what the intensity would have been if there were no grains, the extinction *in magnitudes* is defined by 2.5 $\log_{10}[I_0 (\lambda)/I (\lambda)]$.

The ratio of scattering to absorption is about 2 at visual wavelengths and about 3 at the shortest wavelengths, $1/\lambda$, greater than 8 $(\mu m)^{-1}$ in Figure I 1.1. Near the peak of the extinction curve, $1/\lambda = 4.55$ $(\mu m)^{-1}$, the situation is reversed, however, with absorption as important as scattering.

A curve has been drawn through the data points in Figure I 1.1, a curve which shows what a satisfactory theory of the **interstellar** grains must give for their extinction properties. Moreover, the permissible amount of the grains must match the normalisation value of 1.8 magnitudes per kiloparsec at $\lambda = 0.55$ μm, and the nature of the grains must fit the observed scattering to absorption ratios at the various wavelengths.

Fig. I 1.1.–Interstellar extinction curve for the 3 component microbial model compared to the observational points.

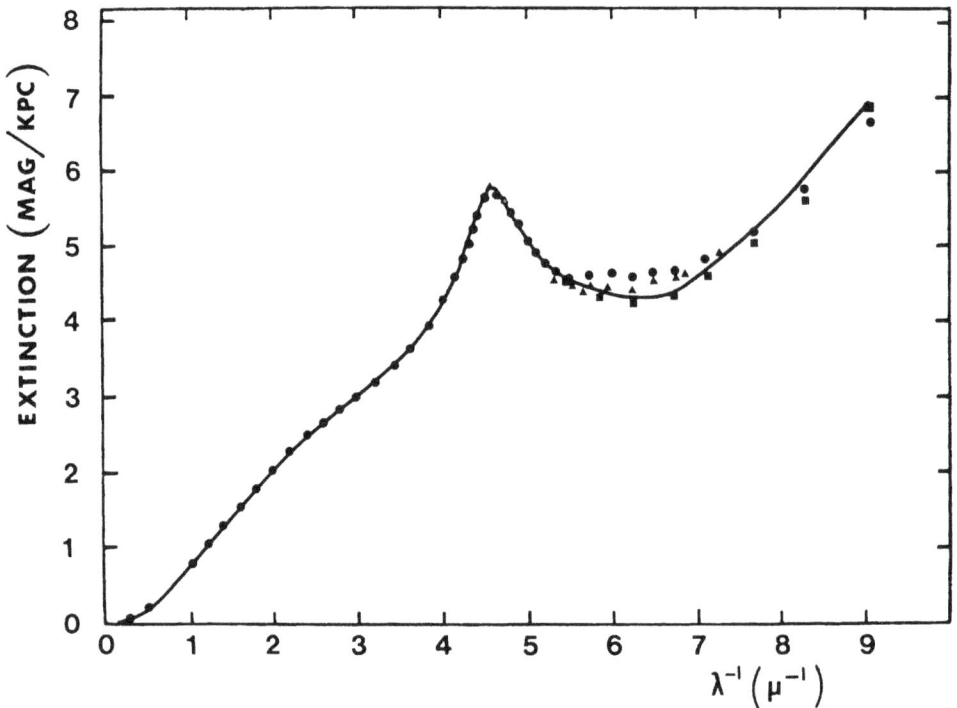

I 2. Proof that Ultraviolet Absorption near $\lambda = 2000$ Å Arises from Small Particles of Free Carbon

The average density of interstellar hydrogen within ~ 1 kpc of the Sun is about 2.10^{24} gm cm^{-3}. Since the cosmic carbon abundance by mass is $\sim 1/200$ of the hydrogen abundance, the total carbon density is $\sim 10^{26}$ gm cm^{-3}, and if we assume the whole of the carbon is in the form of small particles of graphite this value of $\sim 10^{26}$ gm cm^3 will evidently be an upper limit to the density of such particles. Along a column with cross-section 1 cm^2 and length 1 kpc the total mass of small particles of free carbon is therefore not greater than $\sim 3.10^{-5}$ gm. For the particles to produce an extinction by absorption of ~ 6 magnitudes at $\lambda = 2200$ Å (Figure I1.1) the mass absorption coefficient must therefore be at least $\sim 2.10^5$ cm^2 gm^{-1} which turns out to be less by only a modest factor than the largest absorption coefficient that any grain material can have, as will shortly be proved. It follows therefore that no material with appreciably lower cosmic abundance than carbon can be responsible for the ultraviolet absorption of straight. This rules out metallic particles, iron for instance, which not only has an abundance by mass that is $\sim 1/3$ of the carbon, but which has an unfavourable form of refractive index. Whereas graphite as it turns out has a refractive index that yields the maximum possible absorptivity near $\lambda = 0.2$ μm, metallic particles come nowhere near such a condition, as we shall proceed to demonstrate.

The extinction factor due to absorption $Q_{abs}(\lambda)$ is defined to be such that the absorption of energy from an incident light beam of intensity $I(\lambda)\,d\lambda$ (in the wavelength range λ to $\lambda + d\lambda$) produced by a spherical particle of radius a is $Q_{abs} . \pi a^2. I(\lambda)\,d\lambda$, from which definition the mass absorption coefficient of a distribution of such particles can be seen without difficulty to be :

$$\frac{Q_{abs} . \pi a^2}{4\pi a^3 \, \rho/3} = \frac{3}{4}\frac{Q_{abs}}{a\,\rho} \, , \qquad\qquad 12.1$$

Where ρ is the mass density of the material.

A particle with $2\pi a \ll \lambda$ does not attenuate an electromagnetic wave appreciably within itself, and so gives a bigger mass absorption coefficient than a larger particle, which because of its size loses internal efficiency due to a skin-depth effect. To obtain the largest possible absorption coefficient we therefore consider particles for which $2\pi a \ll \lambda$, and then Q_{abs} can easily be calculated from the first order term in the well-known Mie expansion, viz.,

$$Q_{abs} = 24 \frac{\pi a}{\lambda} \frac{2\sigma\lambda/c}{(K+2)^2 + (2\sigma\lambda/c)^2} \qquad\qquad I2.2$$

93

where the real and imaginary parts, K and $2\sigma\lambda/c$, of the complex dielectric constant are given in terms of the complex refractive index, $n-ik$, by the relatins :

$$K = n^2-k^2, \quad 2\sigma\lambda/c = 2nk.$$

It is immediately seen from 12.2 that $Q_{abs} \to 0$ either as $\sigma \to 0$ or as $\sigma \to \infty$, so that absorption of the wavelength λ is small either for dielectric particles of low conductivity σ, or for particles of too high conductivity, for metallic particles.

Substituting 12.2 for Q_{abs} in 12.1, the mass absorption coefficient is :

$$\frac{18\pi}{\lambda\rho} \frac{2\sigma\lambda/c}{(K+2)^2+(2\sigma\lambda/c)^2} , \qquad 12.3$$

the radius a of the particle having disappeared through cancellation, whence it is irrelevant to a calculation of the mass absorption coefficient whether the particles have a single value of the radius or a distribution of values of the radius (only so long as $2\pi a \ll \lambda$ is not violated by an appreciable number of particles in the distribution). It follows immediately from 12.3 that the mass absorption coefficient at wavelength λ is a maximum when the conductivity λ is such that :

$$\frac{2\sigma\lambda}{c} = K + 2 \qquad 12.4$$

In terms of n, k, 12.4 can be written in the form :

$$(n-k)^2 = 2(k^2-1) . \qquad 12.5$$

Figure 12.1 gives experimentally determined values of n, k as functions of the wavelength λ for graphite and for iron. Inspection shows that equation 12.5 is satisfied for graphite at λ a little shortward of 0.2 μm, and at no other wavelength. There is no wavelength for iron at which 12.5 comes at all close to being satisfied, bearing out our previous statement that metallic particles come nowhere near giving maximum absorptivity. Metallic particles fail both for this reason and because of inadequate abundances to explain the observed extinction near $\lambda = 0.2$ μm.

For graphite near $\lambda = 0.2$ μm the mass absorption coefficient is given by substituting 12.4 in 12.3, viz., $18\pi/\lambda\rho(K+2)$, and noting that $2\sigma\lambda/c = 2$ nk and also using 12.3 this is

$$\frac{9\pi}{\lambda\rho} \cdot \frac{1}{nk} \qquad 12.6$$

For graphite, Figure 12.1 gives nk $\simeq$ 1 near $\lambda = 0.2$ μm, and with $\rho = 2.25$ gm cm^{-3} for graphite particles the maximum absorption coefficient (actually attained) is seen from 12.6 to be ~ 6.10^5 cm^2 gm^{-1}. Recalling our previous result that if all the available carbon were in the form of small graphite particles the observed absorption near $\lambda = 0.2$ um would be explained by a coefficient of $\sim 2.10^5$ cm^2 gm^{-1}, we now see that since the actual coefficient is $\sim 6.10^5$ cm^2 gm^{-1}, the observed extinction is explained if about a third of the available carbon is graphite. Two-thirds of the carbon is therefore available for CO gas and for grains with non-absorbing dielectric properties, grains composed of organic materials, for bacteria (entry 15).

Fig. 12.1.–Complex refractive indices of iron and graphite.

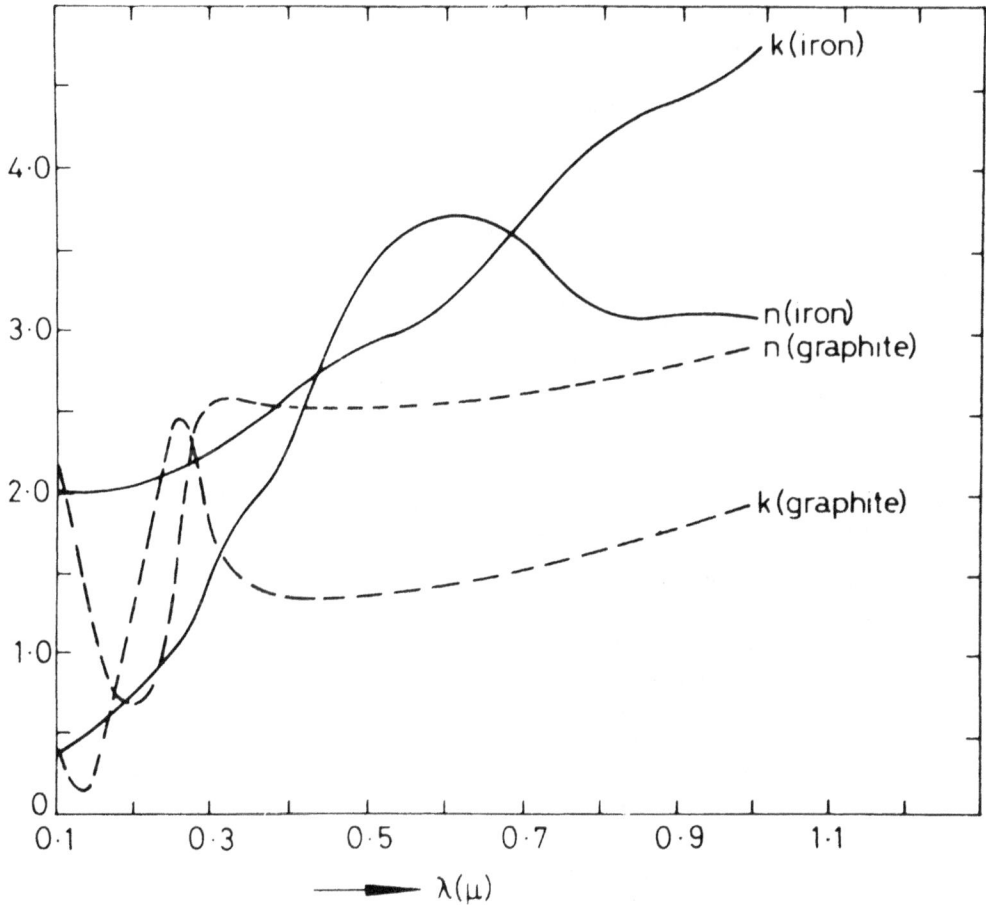

95

An important point remains. Whereas Figure I2.1 gives maximum absorptivity a little shortward of λ = 0.2 μm, the observed maximum extinction occurs a little longward of this, at $\lambda \simeq$ O.22 μm (entry I1). The whole of the present discussion has been for particles with $2\pi a \ll \lambda$, , however. When $2\pi a \simeq \lambda$, higher terms than the first in the Mie expansion must be considered, and these have the effect of moving the wavelength of maximum absorption longward of 0.2 μm. Figure I2.2 shows full Mie calculations for the cases a = 0.01, 0.02, 0.03 and 0.04 μm. A gradual movement of the wavelength of maximum absorption is evident from the four panels of the figure. Inspection shows that it is the calculation for a = 0.02 μm that best fits the wavelength λ = O.22 μm at which maximum extinction is observed (entry I1), indicating that the graphite particles mostly have this particular size.

We are now in a position to appreciate a curious cyclic situation. Microorganisms are particularly sensitive to ultraviolet light with wavelengths in the region of λ =0.22 μm (entry B 3). In the absence of free oxygen at appreciable density, as in interstellar space, microorganisms killed by exposure to ultraviolet light degrade towards a graphitic state thorough a slow loss of water molecules and other volatile products   The graphitic particles thus produced then give optimum shielding against the otherwise lethal ultraviolet. So by sacrificing a fraction of itself a distribution of microorganisms can become self-protecting against further ultraviolet damage, a remarkable provision of nature which scarcely seems accidental.

I 3. Proof the Essentially No Water-Ice Grains are Present in the Distributed Interstellar Medium

Liquid water has a strong absorption band in the infrared at wavelength 2.95 μm, with a measured coefficient of about 12,000 cm^2 gm^{-1}. (For a review of the measurements see W. M. Irvine and J. B. Pollack, *Icarus*, 1968, 324.) The absorption by water-ice is even stronger than that of liquid water with its band centre shifted longward to about 3.1 μm, where the coefficient has been variously measured in the range from 25,000 cm^2 gm^{-1}, to 40,000 cm^2 gm^{-1}, with the lower value given for condensates of amorphous ice below 135K and the higher value for ordinary crystalline ice (A. Leger, J. Klein, S. de Cheveigne, C. Guinat, D. Defourneau and M. Belin, *Astron. Astrophys.*, 79, 1979, 256).

Working with the lower value of 25,000 cm^2 gm^{-1}, the extinction of ice at $\lambda \simeq 3$μm is within a factor 2 of the extinction coefficient of dielectric particles in the visible (the latter does not exceed ~ 50,000 cm^2 gm^{-1}, as seen in entries 14 and 15). If an appreciable fraction of the interstellar grains were water-ice, the measured extinction at $\lambda \simeq 3$ μm should therefore be about a half of the visual extinction, amounting to ~ 1 magnitude for a typical star at a distance of 1 kpc. No such infrared absorption is found.

Fig. I 2.2 Optical efficiency factors for graphite spheres of various radii as functions of $1/\lambda$ (μ −1)

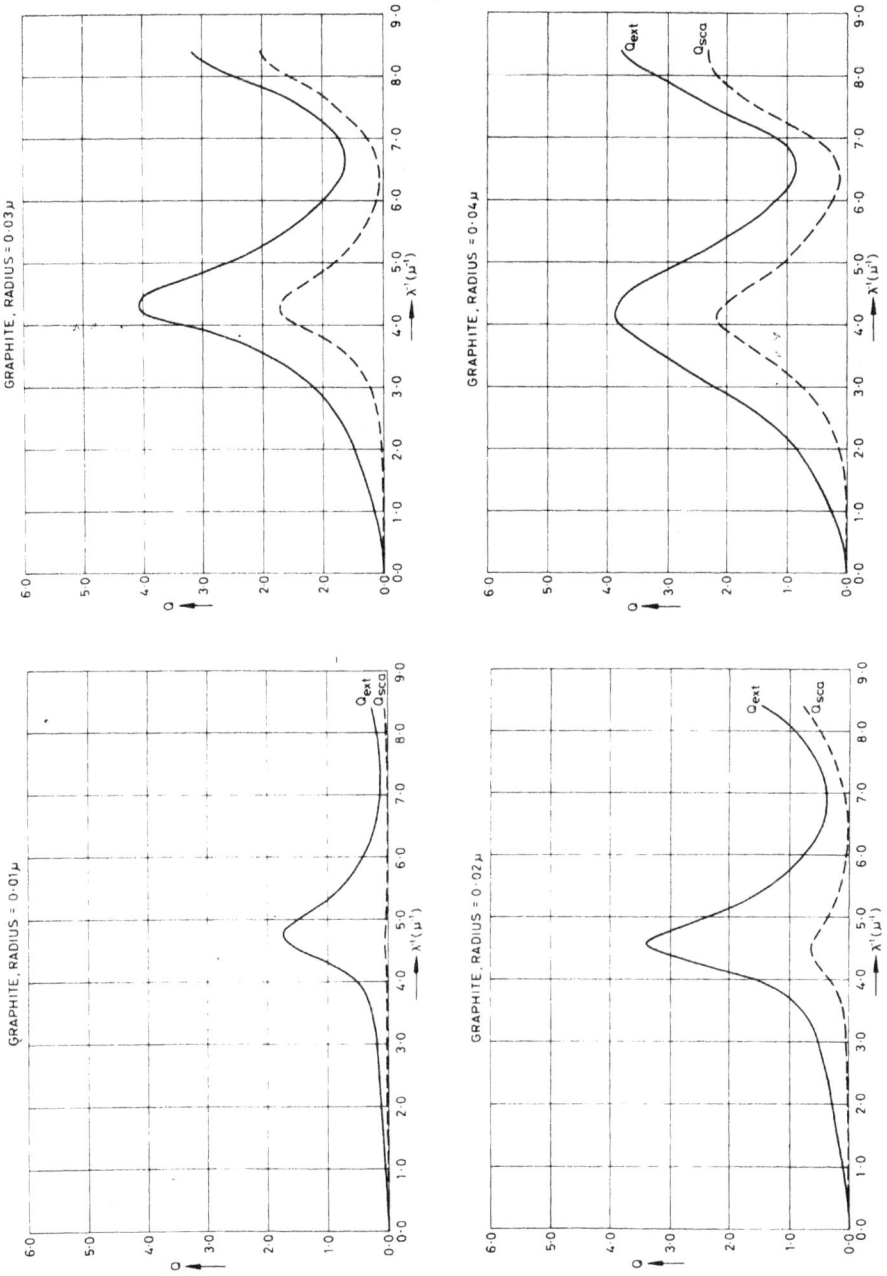

There have been claims of a positive detection of water-ice absorption in the spectra of stars with ~ 10 magnitudes of visual extinction, but only at a doubtful threshold level not greater than 0.1 magnitudes, fifty times weaker than if grains of water-ice were common.

The best test for the presence of water-ice in the distributed interstellar medium comes from recent observations by D. A. Allen and D. T. Wickramasinghe of the infrared source IRS 7 at the galactic **centre**. These authors remark (*Nature.* 294, 1981, 239) :

> We see no evidence for water-ice in the available data. The absorption band at 3.03 μm is displaced from that of water (3.06 μm) and is narrower. If water-ice is present, it contributes little to the absorption. Similarly, solid ammonia (2.91 μm) is not present.

Because of the long path length of ~ 10 kpc from the galactic centre to the Earth, the visual extinction is very large in this case, about 35 magnitudes. Since Allen and Wickremasinghe would readily have seen a water vapour absorption band if it had been as large as 0.1 magnitude, grains of water-ice if present at all cannot be more than half a percent of the total mass of the grains.

Still larger visual extinction values, probably exceeding 100 magnitudes, occur for stars embedded in localised dense clouds, the molecular clouds of which there are about 4000 in our galaxy. Infrared sources exist in some of these clouds, and observations show absorption near 3.1 um amounting in the strongest cases to ~ 2 magnitudes, but more typically about 1 magnitude. If these are indeed evidence of the presence water-ice, the water-ice must still be a small fraction of the total grain mass, no more than a percent or two, otherwise the absorption near 3.1 um would be several tens of magnitudes.

Astronomers have for long been swayed towards thinking that ice would be a material of common occurrence in space, since H_2O is compounded from the commonest molecule. H_2, and after hydrogen and helium, oxygen is the commonest atom. It comes therefore as something of a shock to find the situation otherwise. Ice, if it is present at all, is probably confined to very thin layers condensed on particles of a different nature. Expectations for water-ice were of course based on an inorganic, naturalistic view of the universe. If instead one thinks of a biologically oriented universe, hard-frozen ice is a useless material and the less of it the better. Liquid water is another matter, but water can always be generated in the recycling of microorganisms and so can be made available whenever it is needed.

I 4. Proof that Grains Responsible for Extinction by Scattering are not Mineral Silicates

Some years ago, E. M. Purcell showed how from general principles it was possible to obtain a minimum estimate of the mass of the grains responsible for a total gross amount of extinction equal to an integral taken over the whole wavelength range of the extinction curve (Figure 11.1). This method was then used by Per A. Aannestad and Purcell to show that the grain density needed to explain the integral under the observed extinction curve must at least be a fraction $\rho/200$ of the hydrogen density (*Astrophys. L.*, 158, 1969, 433 ; and *Annual Review of Astronomy and Astrophysics*, 1973, 325). For a mineral silicate $\rho \simeq 3.2$ gm cm^{-3}. The minimum spatial density of such grains necessary to meet Purcell's criterion is therefore $\sim 1.6\%$ of the hydrogen density. Since not more than a third of the extinction can be attributed to the small graphite particles which explain the observed absorption (especially the absorption near $\lambda = 0.22$ µm as in entry 12) mineral silicates would minimally need to have a density of about 1 percent of the hydrogen density in order to explain the scattering of light by the interstellar grains. This minimum requirement is nevertheless about 3 times larger than it is possible for mineral silicates to have. Silicon and magnesium are about equally abundant cosmically, by *numbers* of atoms each about 1/30,000 of the hydrogen abundance. Making up Si and Mg together with O into $MgSiO_3$, the most efficient arrangement, the molecular weight is exactly 100, and the amount by mass of the resulting enstatite grains is 1/3% of the hydrogen density. Since other possible solid dielectric materials among the grains cannot augment the $MgSiO_3$ appreciably, one therefore reaches the conclusion that the available quantity of magnesium silicates is in deficit according to Purcell's minimal criterion by the factor 3.

Here we show, additionally, that this criterion of Purcell would be insufficient even if it could be satisfied, because explaining the integral under the curve of Figure 11.1 is by no means equivalent to explaining the shape of the curve. It is impossible with mineral silicates to account for the smoothly varying extinction at visible wavelengths, which is largely due to a scattering of light (entry 11) so that as well as the mineral-silicate hypothesis being in deficit by gross amount, it is wrong in the detailed form of the extinction curve to which it leads.

Writing $Q_{sca} \cdot \pi a^2 . I(\lambda) \, d\lambda$ for the energy scattered by a spherical particle of radius a out of a light beam of intensity $I(\lambda) \, d\lambda$, the mass absorption coefficient of a distribution of such particles is given (as in equation 12.1 of entry 12) by

$$\frac{Q_{sca} \cdot \pi a^2}{4\pi a^3 \rho/3} = \frac{3}{4} \frac{Q_{sca}}{a\rho} \quad , \qquad 14.1$$

where ρ is the mass density of the material of the particles. (The present restriction to spherical particles is not a serious limitation since corresponding calculations for ellipsoidal particles do not lead to results that are significantly different.) From 14.1, one evidently obtains the highest efficiency of scattering at the particular value of the radius a for which $Q_{sca}(a)/a$ is a maximum.

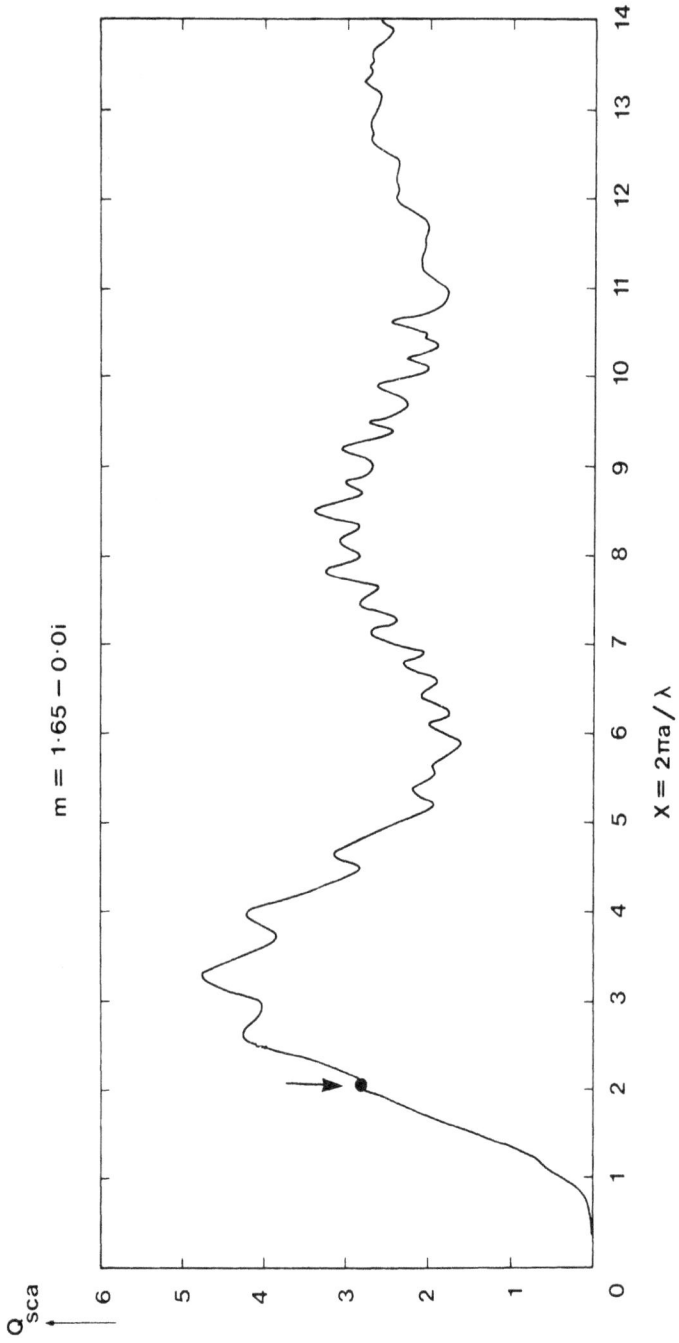

Fig. 14.1 Point indicated by arrow marks position on curve where Q_{sca}/a is a maximum with λ = .55μm.

Figure I4.1 gives Q_{sca} (a) calculated as a function of $x = 2\pi a/\lambda$ for dielectric spherical particles with a real refractive index n = 1.65, the value appropriate for mineral silicates. A highly favourable situation for Q/a occurs if all the particles are close to the point marked in Figure I4.1, where

$$Q_{sca} = 2.8, \quad x = \frac{2\pi a}{\lambda} \simeq 2.1 \qquad\qquad I4.2$$

Such a condition implies an explicit choice of wavelength, which we will take for definiteness to be $\lambda = 0.55$ μm, the wavelength at which Figure I1.1 is normalised. One then has a = 1.8×10^{-5} cm. Putting this value in I4.1 together with Q_{sca} = 2.8, ρ = 3.2 gm cm^{-3} appropriate for mineral silicates, then gives a scattering coefficient of about 35,000 cm^2 gm^{-1}.

It is now easy to calculate the maximum extinction due to silicates for a 1 kpc path length with the hydrogen density taken to be 2.10^{-24} gm cm^{-3}, its usually assigned value for the solar neighbourhood. Thus the mass of hydrogen in a column of 1 cm^2 cross-section with a path length of a kpc is ~ 6.10^{-3}gm. Since the mineral silicates have at most a mass equal to ~ 1/300 of the hydrogen, the mass of mineral silicates available along a 1 kpc path length is no more than ~ 2.10^{-5}gm. Hence with an upper limit to the scattering coefficient of about 35,000 cm^2 gm^{-1}the maximum optical depth is about .75, giving ~ 3/4 magnitude of extinction. This is at $\lambda = 0.55$ um where the observed extinction is taken to be 1.8 magnitude (Aannestad and Purcell use a normalisation of 2 magnitudes) nearly three times larger than the extinction just calculated. The discrepancy can be reduced somewhat at this one particular wavelength by noting that only some two-thirds of the oberved extinction at $\lambda = 0.55$ μm arises from scattering (entry I1). The situation is then somewhat more favourable than the situation considered by Aannestad and Purcell, precisely because the wavelength is fixed here, whereas the latter authors consider the extinction problem with respect to the whole range of wavelength. Indeed this slight improvement at $\lambda = 0.55$ μm has been bought at a terrible price. The parameter $x = 2\pi a/\lambda$ changes for particles of fixed radius a as the wavelength changes, with the point of Figure I4.1 sliding up the curve as the wavelength decreases, and down the curve as the wavelength increases. In effect, Figure I4.1 becomes the extinction curve, disastrously, because the observations of Figure I1.1 show nothing of the oscillations and of the overturn that develop as x increase beyond 2.5.

What one attempts to do now is to choose a particle distribution N (a) da, where N (a) is the number of particles with radii between a and a + da, the aim being to adjust N (a) in order to smooth the oscillations of Figure I 4.1 at x > 2.5 as much as possible. Such a procedure, however, has the immediate effect of decreasing the mass absorption coefficient appropriate to the whole particle distribution, and the discrepancy of abundance discussed above is soon increased back to the factor 3 obtained from the integral method of Aannestad and Purcell. Besides which, this procedure turns out to be a counsel of despair, for no function N (a) that we have ever been able to find in long-sustained calculations of this kind has produced agreement with the smooth variation of the observational points of Figure I 1.1. In our experience the oscillations of Figure I 4.1 at x >2.5 always cause serious trouble. Claims to the contrary have appeared in the literature, but in no case of which we are aware has such a claim been accompanied by a visual correspondence of the claimed calculations with the observational points (in a plot of the kind shown in Figure I1.1 which has been the standard form of presentation of data since the 1930's). Nor have such claims been accompanied by a precise statement of the parameters used by the authors, so that one is not permitted to repeat the calculations in order to check the claims. In these circumstances we remain sceptical of such statements.

I 5.   Proof the Grains Responsible for Extinction by Scattering are Bacteria Which May or May Not be Viable

An intercomparison of Figures I1.1, I4.1 and I5.1 goes to the heart of the problem of the nature of the dielectric grains responsible for the extinction by scattering of visible starlight. Changing the real value of the refractive index from n = 1.65 (appropriate for mineral silicates) to n = 1.167 appropriate for rod-shaped bacteria that have lost their free-water, yields a smooth curve for Q_{sca} from which the unwanted oscillations of Figure I 4.1 have disappeared, so that now we have a chance to explain the smoothness of the data points in Figure I1.1.

Before actually calculating the extinction curve, let us derive n = 1.167 for dried bacteria. About two-thirds of the volume of a bacterium is occupied by free water under normal conditions here on the Earth. With n = 1.5 for the one-third of a bacterium that is of biochemical composition, and with n = 1.33 for the two-thirds that is water, the average refractive index for a water-filled bacterium would be

$$\frac{1}{3} \ (1.5 + 2 \times 1.33) \ = \ 1.39 \qquad\qquad 15.1$$

It is of interest to note in passing that particles in the atmosphere of Jupiter have been shown to be of a bacterial size and to have n $\simeq$ 1.38 (D. L. Coffeen and J. E. Hansen in *Planets, Stars and Nebulae Studied with Photopolarimetry*, ed. T. Gehrels, University of Arizona Press, 1974). When a bacterium dries out, however, as it would do in space, the cell wall being exceedingly strong does not collapse inward. It remains in position and cavities develop inside the bacterium, cavities occupying about two-thirds of the volume. Since the refractive index of free space is unity, the mean refractive index is then

$$\frac{1}{3} \ (1.5 + 2) \ = \ 1.167 \qquad\qquad 15.2$$

the value for which Figure I 5.1 has been calculated. (A small particle with a non-uniform interior behaves in first approximation like a uniform particle of the same outer boundary having the mean refractive index of the actual irregular particle.)

By using the size distribution of bacteria given in entry B5 the extinction due to scattering can now be calculated without any further information being needed. At a particular wavelength each size-bin of the histogram of Figure B5.1 has an easily determinable value of the parameter $x = 2\pi a/\lambda$. . The value of Q_{sca} at this value of x is read-off from Figure I 5.1 and is then multiplied by the height of the appropriate component of the histogram. This is done for each size-bin and the results are added. The same procedure is carried-out at other wavelengths, so yielding a catalogue of relative opacity values as a function of λ. The catalogue is finally normalised to yield opacity values $\tau(\lambda)$ such that $\exp(-\tau(\lambda))$ has some assigned extinction value at a particular λ, say at $\lambda = 0.55$ μm. The result of this procedure is shown in Figure I 5.2, the data points being those of K. Nandy (*Publ. Roy. Obs. Edin.*, 3, 1964, No. 6). Combining this **execellent result obtained without any free parameter at all (since Figure B5.1 was** obtained from laboratory data), with the absorption due to small graphite particles discussed in entry I2, one obtains the curve drawn in Figure I1.1, a curve that passes impressively through the averaged data points given by Sapar and Kuusik (reference in entry I1).

Most of the particles in the size-distribution of Figure B5.1 fall at visual wavelengths on the rising linear part of Figure I5.1. On this linear part $Q_{sca}/x \approx 1/3$, $x = 2\pi a/\lambda$. Inserting $Q_{sca}(a)/a = 2\pi/3\lambda$ in I4.1 gives $\pi/2\lambda\rho$ for the mass scattering coefficient, which for $\lambda = 0.55$ μm and $\rho = 0.5$ gm cm^3 – appropriate for bacteria with cavities occupying two-thirds of their interior – is about 50,000 cm² gm^1 larger than the value $\sim$ 35,000 cm² gm^1 calculated in entry I 4 for mineral silicate particles. Since the value of $\sim$ 35,000 cm² gm^1 led to a required mass density of grains that was about 1 per cent of the hydrogen density, required that is to say to explain the observed extinction due to scattering, a similar calculation for the larger coefficient of 50,000 cm² gm^1 would lead to a required grain density equal to about 0.7 per cent of the hydrogen density.

In material with the usually-accepted cosmic abundances of the elements the carbon mass density is ˜ 0.45 per cent of the hydrogen density. Of this, the graphite particles discussed in entry I 2 contribute about one-third so that the remaining carbon amounts to 0.3 per cent of the hydrogen density. This remining carbon is available for bacteria*. When

---

*P. M. Solomon and D. B. Sanders in *Giant Molecular Clouds in the Galaxy*, Pergamon 1980, page 44, give as little as 1/30 of the carbon observed as CO, and the amount of carbon present in gaseous organic compounds is much smaller still. Very little carbon is needed therefore for these gases.

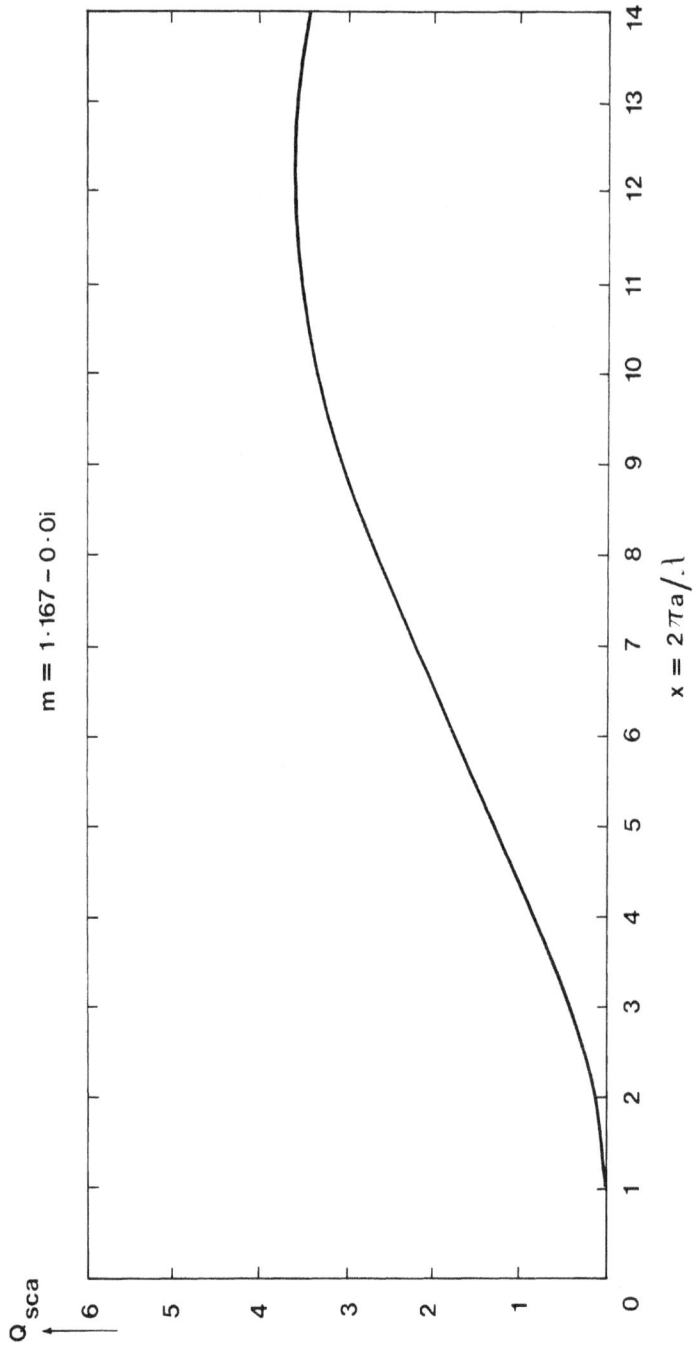

Fig. 15.1.

m = 1·167 − 0·0i

Q_{sca}

$x = 2\pi a/\lambda$

104

water is excluded, about forty per cent of biological material is carbon. Hence the amount of available carbon is sufficient when taken-up in biological material to yield ~ 0.75 per cent of the hydrogen density, almost precisely as required, unlike the large discrepancy found for mineral silicates in entry I 4.

It would be strange if the generally-similar grains observed in other galaxies were different in their nature from those of our own galaxy. The widespread occurrence of grains in other galaxies therefore suggests that microorganisms exist in profusion throughout the universe, and hence that creatures like ourselves compounded from genetic fragments and microorganisms by the evolutionary processes discussed in entries E are almost surely widespread throughout the universe. To some, this may seem like an absurd conclusion but absurdity, at least in this case, lies in the eye of the beholder. As the outcome of our modern helter-skelter of an educational process we emerge with Macbeth's concept of a universe 'full of sound and fury, signifying nothing' almost irreversibly programmed into our brains. The facts tell a different story. The facts clearly – one would think overwhelmingly – tell of a biologically-oriented universe, if not indeed of a biologically-controlled universe.

I 6.   Further Proof That Interstellar Grains are Biological in Origin

A program for obtaining well-calibrated infrared spectra with good instrumental resolution was set-up about two years ago by A. H. Olavesen at University College, Cardiff, and has been used by S. Al-Mufti to examine a wide variety of materials—biological, organic but not biological, and inorganic. Figure 16.1 is an example of the results, the spectrum of dried *E. coli* measured at a temperature of 350°C to 400^{b}C.

An unexpected discovery emerged from an intercomparison of the many spectra obtained in this program. The form of the spectra near 3.4 µm was very nearly invariant for biological materials, being largely unaffected either by the temperature or the nature of the specimen—only so long as it was biological. Three examples normalised together at 3.4 µm are shown in Figure I6.2. Non-biological materials, on the other hand, while showing absorption near 3.4 µm if they were organic (due-to the stretching mode of C-H bonds) never reproduced the distinctive signature of the biological specimens. Nor do any non-biological materials with infrared spectra published in the literature, as far as we are aware.

When observational results for the astronomical infrared source GC-IRS7 were obtained by D. A. Allen and D. T. Wickremasinghe (*Nature*, 294, 1981, 239) a comparison of this laboratory-derived signature at 3.4 µm with the data for IRS7 was made immediately, with the outcome shown in Figure 16.3, a result that we consider as thoroughly convincing proof of the biological nature of the interstellar grains. (Note the expanded linear scale of the ordinate and therefore the closeness of the fit of the points to the curve.)

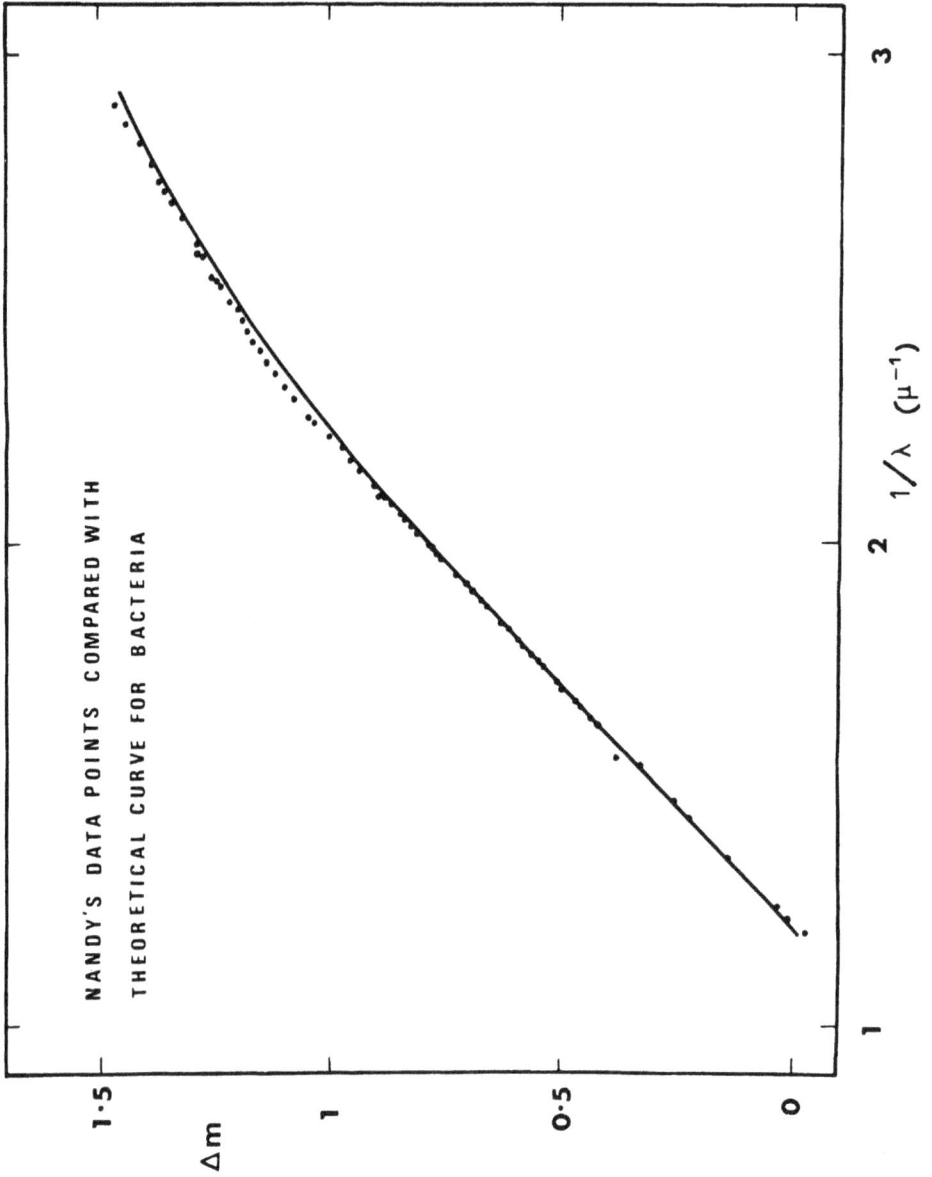

Fig. 15.2. Extinction calculated for bacterial size distribution of grains with n = 1.167 − 0.0i compared with data due to Nandy.

Figure 16.4 shows the full wavelength range of the astronomical data, together with a smooth upper curve which Allen and Wickremasinghe consider to be how the continuum of the underlying infrared source would appear if there were no foreground absorption at various wavelengths. The shape of this smooth continuum is very like a black-body curve for a temperature of about 1100K, which black-body form was used in the simple calculation that converted the invariant laboratory results at ~ 3.4 μm into the curve shown in Figure 16.3.

Several further interesting results can be obtained. In Figure 16.4 the minimum of the observational point near 3.4 μm has an ordinate value of about 8.5, whereas the continuum at 3.4 μm has an ordinate value of about 12.5. From these numbers the extinction at 3.4 μm is 0.42 magnitudes*. We ourselves would prefer to lift the continuum moderately above the curve of Figure 16.4 (while retaining its shape) because we do not think the absorption longward of 3.6 μm, while small, is strictly zero. Such a lifting of the continuum should increase the extinction at 3.4 μm moderately, say by a third to 0.56 magnitudes. The interesting results in question emerge from comparing this extinction value (or some similar value, depending on how one sets the continuum of the underlying infrared source) with Al-Mufti's laboratory determination of ~ 750 cm² gm¹ for the absorption coefficient of biological material at 3.4 μm*. In entry 15 the visual extinction coefficient was found to be ~ 50,000 cm³ gm¹, about 200/3 times larger than Al-Mufti's value at 3.4 μm. So with 0.56 magnitudes of extinction at 3.4 μm the visual extinction should be ≅ (200/3) (0.56) ≅ 37 magnitudes ; close to that estimated by other criteria (D. A. Allen and D. T. Wickramasinghe, *Nature*, 287, 1980, 518).

The mass required to produce 0.56 magnitudes of extinction for a material with an absorption **coefficient** of 750 cm² gm⁴ is ~ 7.6× 10⁻⁴ gm present in a column of cross-section 1 cm² extending from the Earth to the source, in the case of GC-IRS7 a path-length of ≅ 3.10²²cm, about 10 kpc.

The required average grain density along the line of sight is therefore ~ 2.5 × 10²⁶gm per cm³ Now in entry 15 the grain density is shown to be about three-quarters of a percent of the hydrogen density. Hence one deduces that the average hydrogen density along the line of sight is ~ (400/3).(2.5 × 10²⁶) ≅ 3.3 × 10²⁴gm per cm³, or about 2 hydrogen atoms per cm³. P. M. Solomon (private communication) informs us that his personal estimate for this average to the galatic centre is indeed 2 hydrogen atoms per cm³, although some of his colleagues prefer the value of 1.2 atoms per cm³, the same as for the solar neighbourhood.

---

*2.5 $\log_{10}$ (12.5/8.5) = 0.42.

---

*From Figure 16.1, $\bar{e}^\tau$ = 0.57 at λ =3.4 um, ie., τ = 0.56 According to Al-Mufti, the quantity of biological material contained in a KBr disk of diameter 1.3 cm required to give this value of r is close to 1 milligram. This gives 560 π (0.65)² = 750 cm² gm¹ for the absorption coefficient.

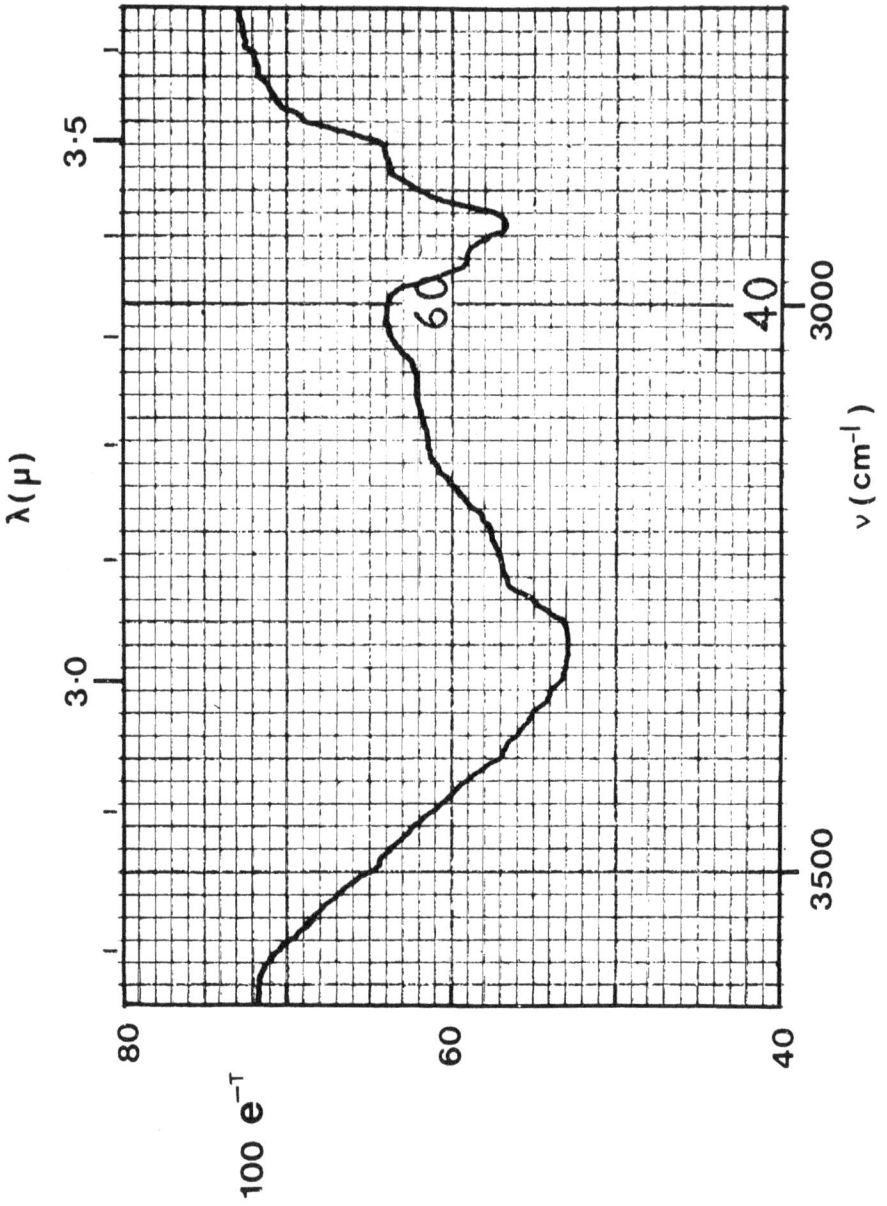

Fig. I 6.1. Enlarged labortory spectrum over the 2.6–3.6 um region for *E-coli* heated to 350°C from which transmittance values were measured for the calculation in Fig. 2.

108

Fig. I 6.2. The measured transmittance curves of micro-organisms. For *E-coli* a dry mass of 1.5 mg was dispersed in a KBr disc of radius 0.65 cm.. The trnsmittnce data for yeast was normalised to agree with the *E-coli* curves at at $\lambda = 3.40 \mu m$.

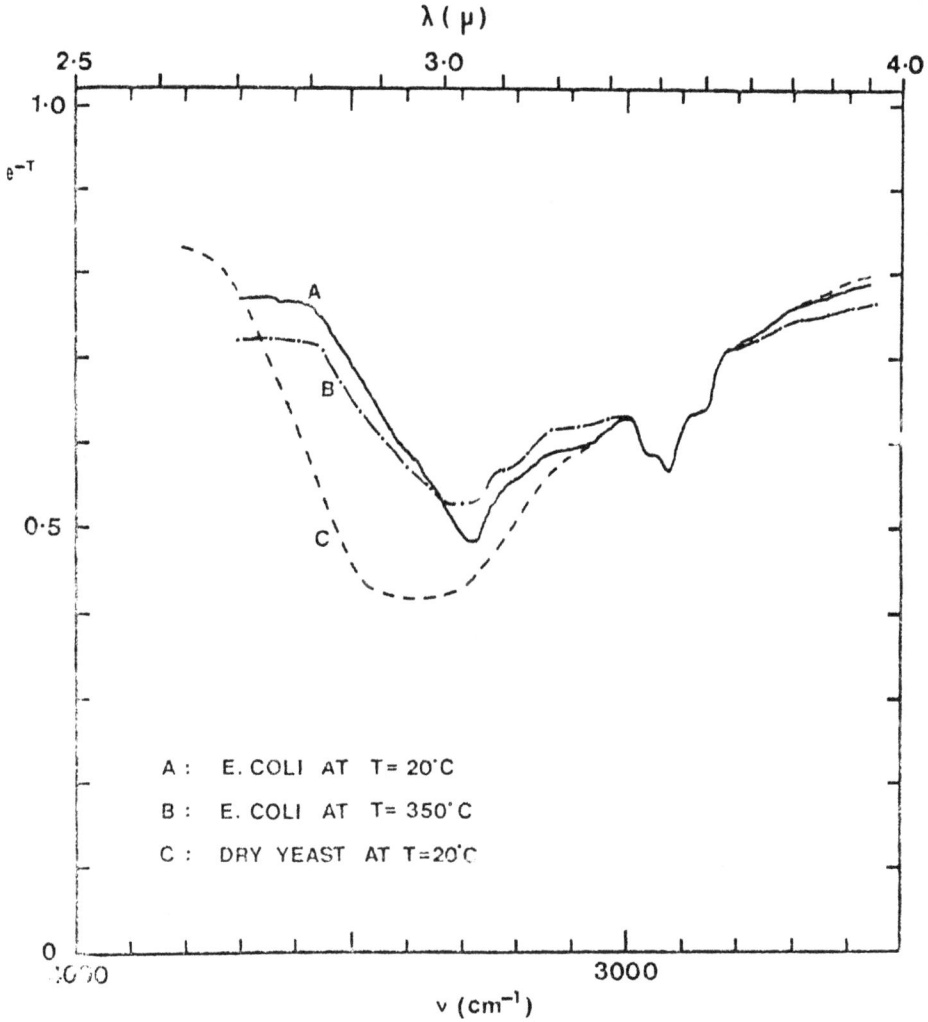

A : E. COLI AT T= 20°C

B : E. COLI AT T= 350°C

C : DRY YEAST AT T=20°C

109

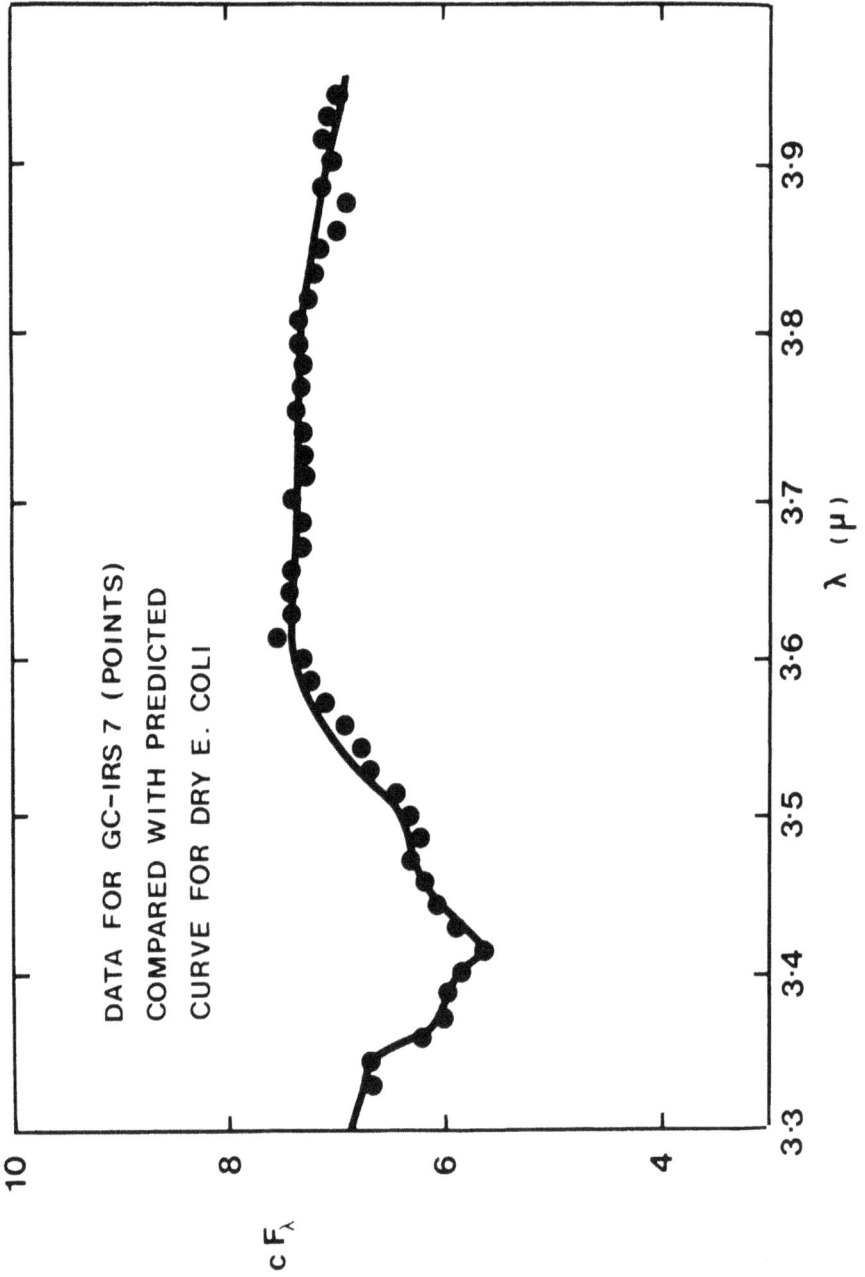

Fig. 16.3. The waveband 3.3-3.9 um showing the detailed agreement between bacteria and IRS 7. We note that the flux is on a linear scale, so that the maximum departures seen here are no more than a couple of percent. The size of the points represents the extent of the estimated errors in both wavelength and flux.

DATA FOR GC–IRS 7 (POINTS)
COMPARED WITH PREDICTED
CURVE FOR DRY E. COLI

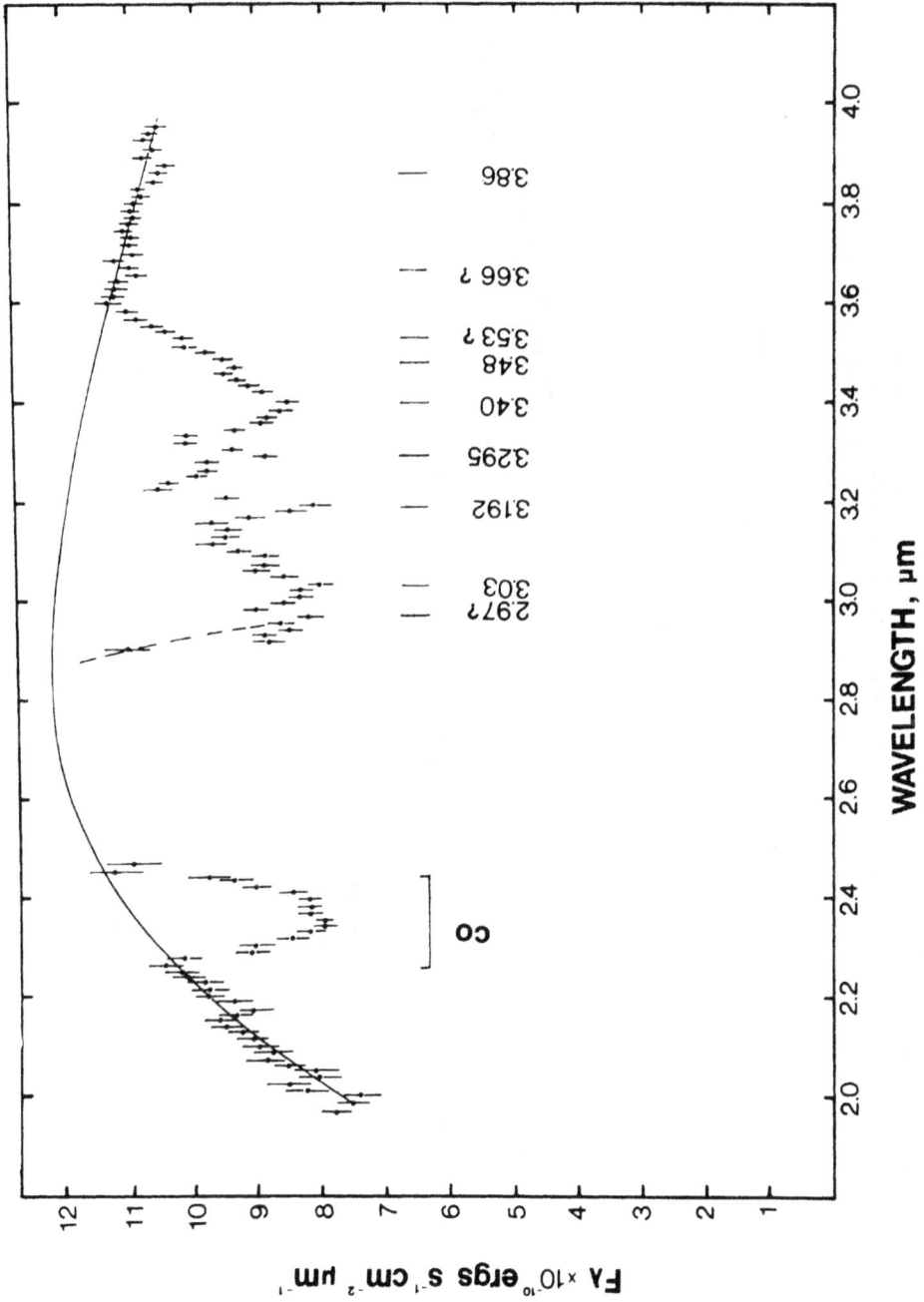

Fig. I 6.4. Observed flux curve for GC-IRS 7 (Allen and D. T. Wickremasinghe, 1981).

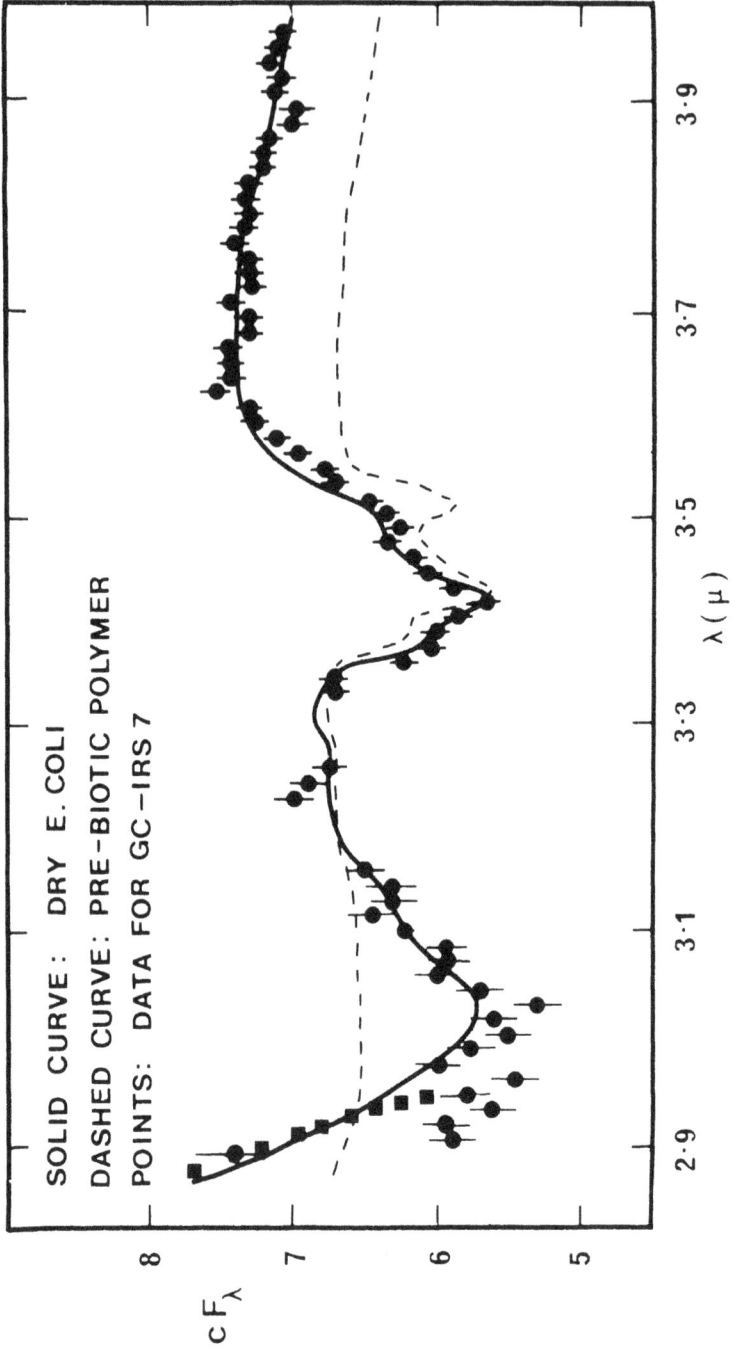

Fig. 16.5   The observed relative fluxes for GC–IRS 7 compared with two models. The circles represent data for ISR 7 obtained in May 1981. The curved line defined by the filled squares represents data over the shortest infrared wavelenghts obtained in July 1980. The solid curve is the predicted curve for bacteria (heated to 350°C), the dashed curve is the least bad fit for a prebiotic polymetic mixture obtained by ultraviolet irradiation in a reducing atmosphere.

SOLID CURVE :   DRY E. COLI

DASHED CURVE: PRE-BIOTIC POLYMER

POINTS: DATA FOR GC –IRS 7

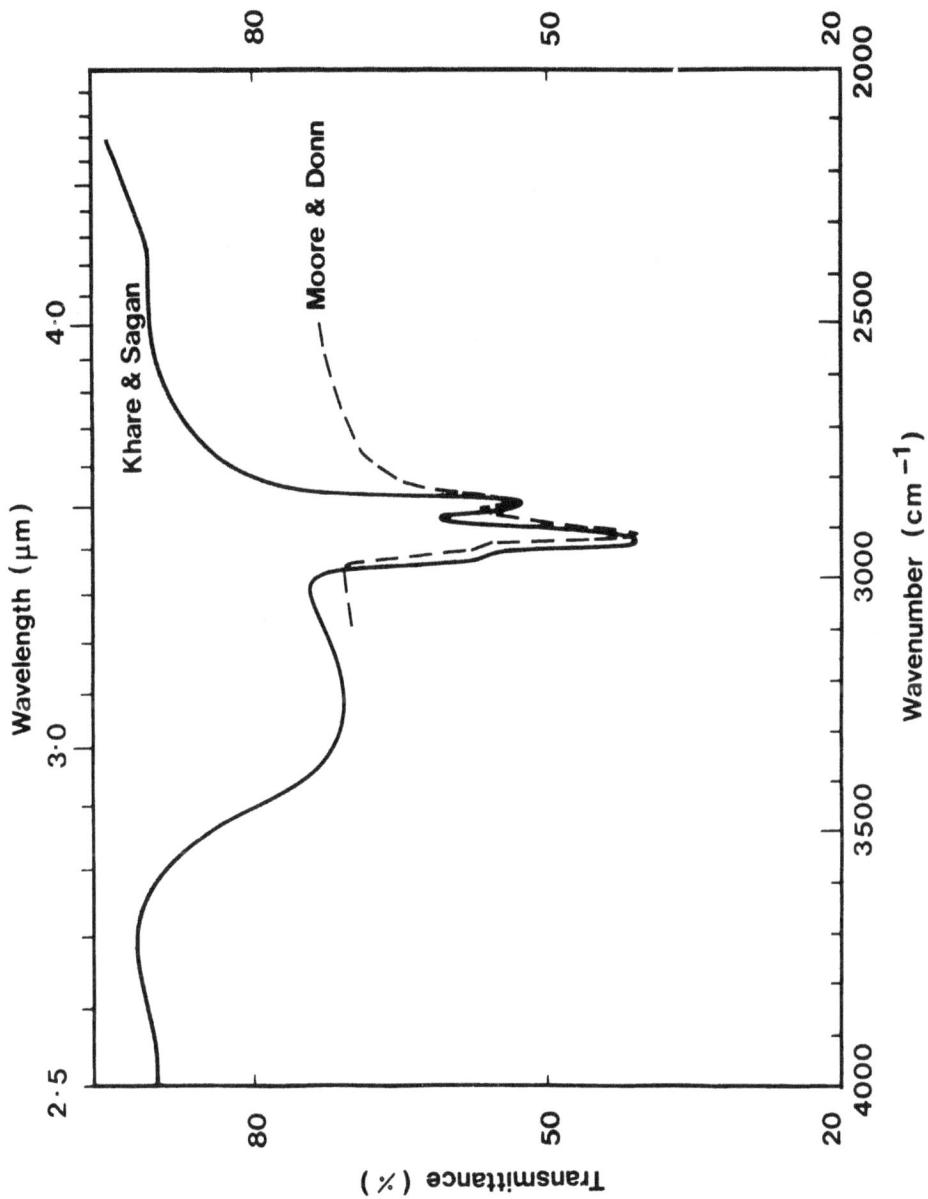

Fig. 16.6. Comparison of transmittnce curves for proon irradicated residue (Moore and Donn) with UV irradiated residue (Khare and Sagan).

113

We have searched the literature for non-biological materials that might give a tolerable approximation to the astronomical data but have found none that is explicitly obtainable. The least bad fit was a polymeric material obtained by irradiating a reducing atmosphere of inorganic gases, but this least bad material gave only the poor correspondnce with the data shown by the broken curve of Figure 16.5 (for a description of how to obtain such a polymeric material see B. N. Khare and C. Sagan, *Icarus*, 20, 1973, 311 ; and for what its spectrum is like see R. F. Knacke, *Nature*, 269, 1977, 132).

A very similar spectrum has recently been obtained by M. H. Moore and B. Donn (*Astrophys. J.* 257, 1982, L47), the two being shown together in Figure 16.6 Whereas Khare and Sagan irradiated a mixture of reduced inorganic gases with ultraviolet light, Moore and Donn exposed a solid mixture of reduced inorganics to intense particle irradiation. The interesting conclusion to be drawn from this comparison is that any experiment which begins from an intimate mixture of reduced inorganic molecules, say, CH_4 NH_3 and OH_2, and proceeds in any way to disrupt the robust bonds in these molecules, will yield essentially the same end products when the disruptive agent is removed, with the same atoms going through essentially the same network of chemical reactions. For this reaon no synthetic laboratory spectrum obtained by such a procedure will yield a comparison with the astronomical data that is any better than the broken curve of figure 16.5.

I 7. The Silicate Band at 10 μm

Interstellar grains emit and absorb radiation over a broad band centred at ~ 10 μm, as was first shown observationally by N. J. Woolf and E. P. Ney (*Astrophys. J.*, 155, 1969, L185 ; and also E. P. Ney and D. A. Allen, *Astrophys. J.*, 155, 1969, L193). Since materials containing many Si - O and Si - O - Si inkages have particularly strong absorptions at these wavelengths, and since evidence at shorter wavelengths (entries I 5 and I 6) demonstrates the biological nature of the interstellar grains, it is natural to recall the class of microorganisms which have Si-bonds appearing importantly in their cell structures. Polysaccharides also have appreciable absorption in the 10 μm wavelength region, and polysaccharides are also a significant component of the cell walls of such microorganisms.

Figures I 7.1 and I 7.2 answer the question of how far silicious microorganisms can explain the astronomical data for the silicate band at ~ 10 um. The points of Figure I 7.1 are observations of infrared emission by grains in the Trapezium region of the Orion nebula, grains that are heated to temperatures in the range 150K to 200K by the hot Trapezium stars (W. J. Forest, F. C. Gillett and W. A. Stein, *Astrophys. J.*, 192, 1975, 351 ; and the same authors in *Astrophys. J.*, 195, 1975, 423). Numerous investigations have shown that grains with the emission properties of Figure I 7.1 also explain the observations for many other objects, notably for cases where cool grains on the near-side of a source of infrared radiation act in absorption instead of in emission. The

Fig. I 7.1.   Observed flux from the Trapezium Nebula (Forest *et al* 1975 a,b) compared with the behaviour of grains with opacity values as in Figures I 7.2 and with a tmperature of 175K. Normalisation of curve is to F_λ =6 × 10 $^{-16}$ Wcm^{-2} s^{-1} μm^{-1}at λ = 9.5μm.

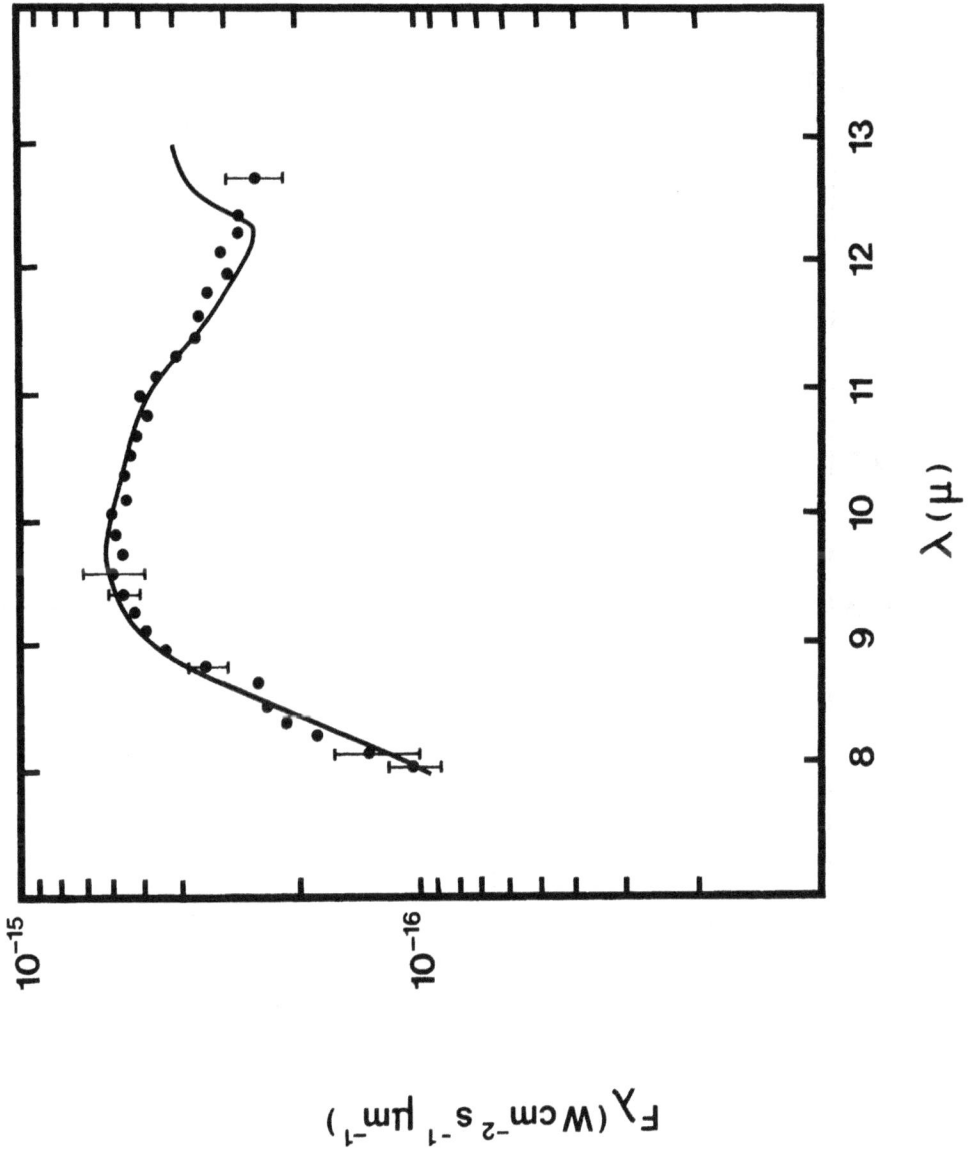

Fig. I 7.2.   infrared transmittance of diatoms for $\lambda \gtrsim 8\,\mu m$ compared with the best fitting values (Table 1) for matching the Trapezium data.

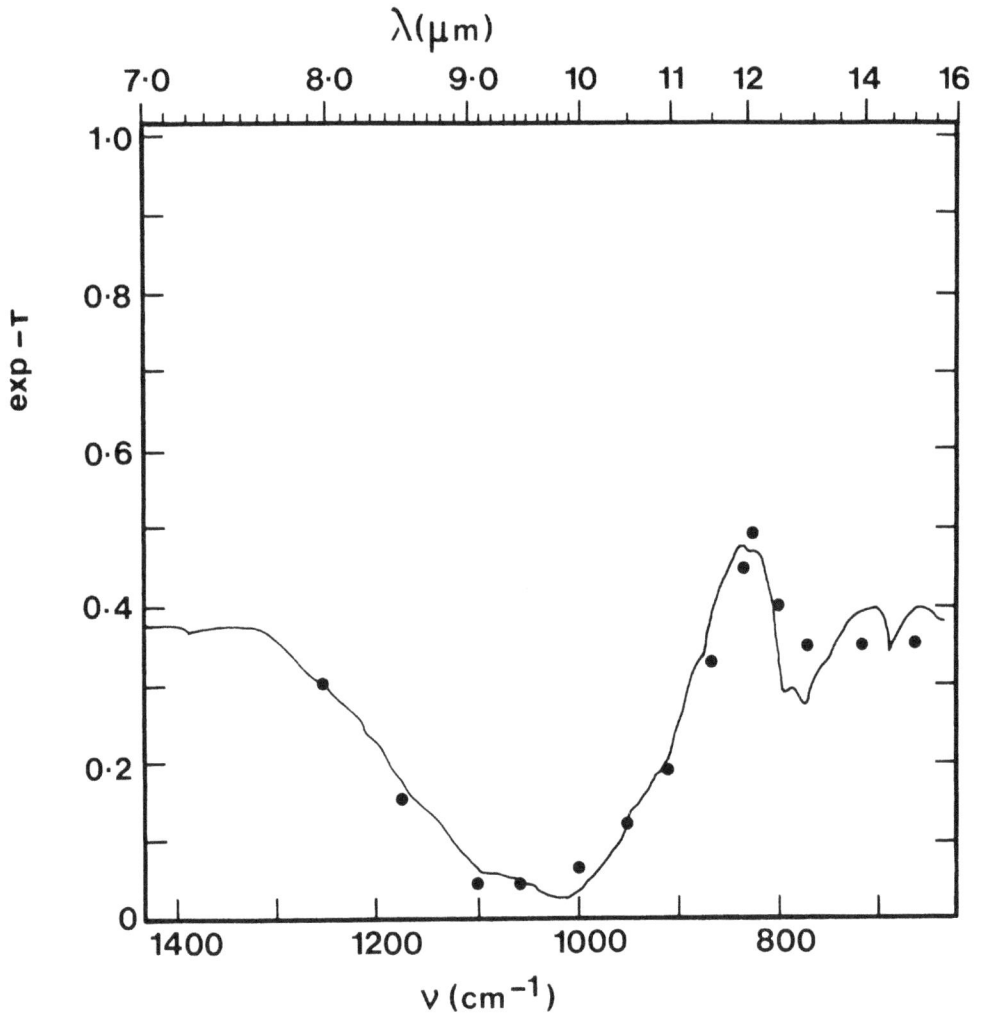

grains of the Trapezium region, 'Trapezium material' as they are often called, typify the behaviour in the silicate band of grains throughout the galaxy, and so an understanding of 'Trapezium material' is regarded as the key to understanding the longwave properties of grains generally.

The curve of Figure I 7.1 shows the calculated emission from grains at a temperature of 175K, assuming the grains to have the transmittance properties defined by the points in Figure I 7.2, while the curve of Figure I 7.2 is the measured transmittance of a heterogeneous distribution of silicious organisms cultured from a bucket of local river water, with the point at 8 μm normalised to fall on the laboratory spectrum. Pure cultures of particular silicious organisms differ only slightly one to another except near 13 μm, where the differences are somewhat larger, for which reason it is thought sufficient that the required points (●) match the laboratory spectrum near 13 μm only in a general way. The match in the rest of the wavelength range is good. To appreciate how good, consider the measured transmittance curve for olivine particles shown in Figure I 7.3 (from K. L. Day, *Astrophys. J.*, 199, 1975, 660). Attempts have been made to ameliorate the jagged structure of olivine near 10 μm – hopeless for explaining the smooth gradually-varying observational points of Figure I 7.1 – by adding effects from other minerals such as dunite and plagioclase. This procedure is unacceptable, however, because it brings in elements such as Na, Al, K and Ca, which are of significantly lower cosmic abundances than the Mg, Si, Fe and O present in olivines. The latter are already lacking in abundance by a considerable factor (entry I3) and a further reduction by introducing elements (not in trace quantities but as principal constituents of minerals) of still lower cosmic abundances by as much as a whole order of magnitude is clearly ruled-out.

Grains of SiO_2 have two very strong bands at ~ 8.7 μm and ~ 12.7 μm (R. F. Knacke, *Nature,* 217, 1968, 44 ; and P. G. Martin, *Astrophys. Letter.,* 7, 1971, 193) both with central absorption coefficients comparable to water-ice at ~3.1 um (entry I3), and like the water-ice absorption neither is seen observationally. Like water-ice, quartz grains must either be wholly absent or very nearly absent from the interstellar medium, a result that we would find astonishing if the interstellar grains were of inorganic origin–it seems inconceivable that circumstances would always conspire to exclude quartz, one of the most refractory of minerals. It is the silicious organisms of biology which work to exclude quartz, by preying on SiO_2 as a nutrient. The observed absence of quartz grains from the interstellar medium is again suggestive of a biologically oriented universe, perhaps indeed of a biologically-controlled universe.

Wheres graphite is the ultimate degeneration product of wholly carbonaceous organisms (entry I2) silicon carbide would be expected to appear in the degeneration of silicious organisms. Silicon carbide is believed to exist in some circumstellar envelopes. Grains in circumstellar envelopes are usually thought of as condensates occurring in gaseous outflows from the underlying stars, a natural-seeming view if one believes the universe to be inorganically-oriented. If one believes in a biologically-oriented universe, on the other hand, the sensible-seeming view would be that circumstellar envelopes are the products of degeneration of microorganisms, destroyed by the radiation of the underlying stars.

Fig. I 7.3.   Laboratory measurement of the spectrum of silicates (K. L. Day, *Astrophys. J.*, 199, 1975, 660).

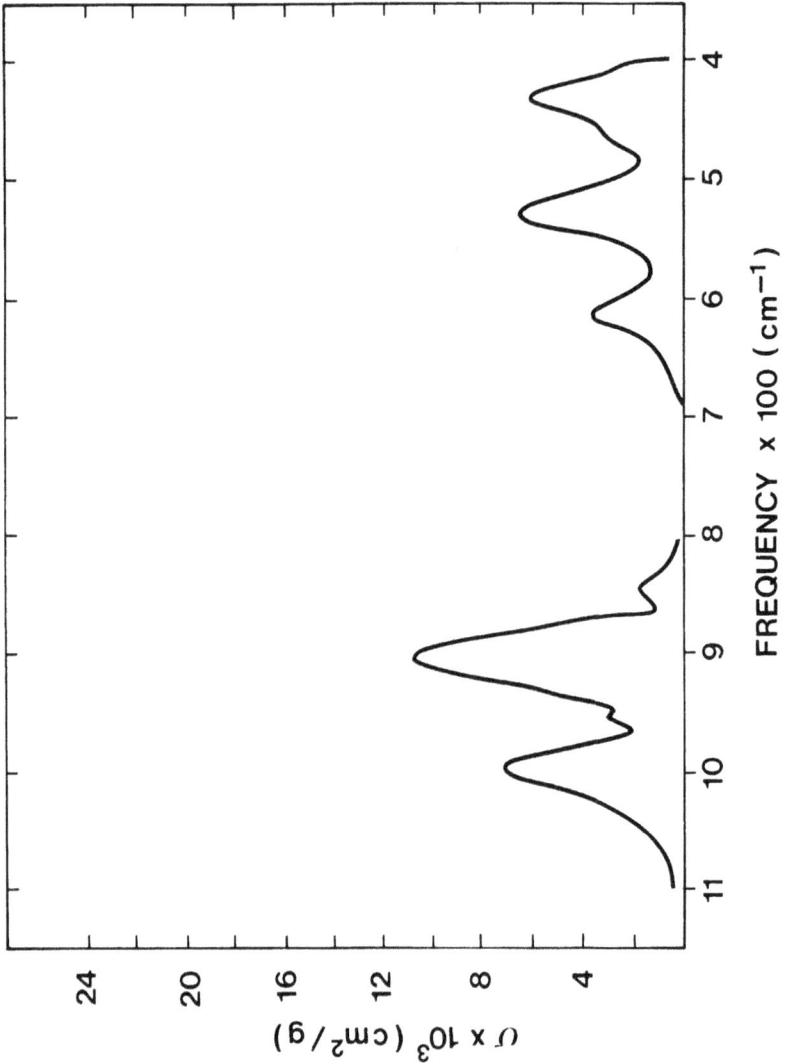

M

Meteorites

M 1. Proof that Life Antedated the Earth

Carbonaceous meteorites are regarded as being the most 'primitive' material available to us. Excluding elements with volatile chemical compounds, their composition with respect to the abundances of the elements is remarkably similar to the Sun, which is considered to reflect the composition of the cloud of interstellar gas in which our solar system condensed some 4.6×10^9 years ago. Although the carbonaceous meteorites contain individual grains which appear to have been melted, the meteorites have not been melted as a whole – indeed, the matrix material which surrounds mineral grains and the so-called chondrules is thought in some cases never to have been heated to a temperature above about 150°C. Radioactive data for individual grains give ages of $\sim 4.6 \times 10^9$ years, indicating that the process of aggregation of the carbonaceous meteorites took place in the very early history of the solar system, probably even before the Earth was formed.

These meteorites are believed to be derived from cometary objects that experienced the process of 'rounding-up' discussed in entry C4, whereby a comet initially of long period has its period and aphelion distance reduced, the latter eventually coming within the radius of the orbit of Jupiter, with the plane of the comet's motion around the Sun developing a low inclination to the plane of Jupiter's orbit. Such a dynamical history, which can take place at times long after the formation of the solar system itself, exposes a cometary body to collision with one or other of the many asteroids that move around the Sun in more or less circular orbits mostly in the region between Mars and Jupiter. When collisions occur, bits are struck-off both comets and asteroids, and the meteorites are these bits. Hence meteorites as a whole are a complex mixture of cometary material and asteroidal material, with the carbonaceous meteorites coming from the comets.

As their name implies, the carbonaceous meteorites contain carbon, with concentrations upwards of 2% by mass. Partly from their carbon concentrations, the presence or otherwise of chondrules, the oxidation or otherwise of sulphur, whether the iron content is magnetic or not, these meteorites have been divided into classes, Types I, II and III, with Type I viewed as the most 'primitive', and Type II the next most primitive. The greater part of the carbon in Types I and II is present as a compound of very high molecular weight, insoluble in all solvents –the so-called insoluble residue, a compound found by chemical analysis to contain hydrogen, oxygen and probably nitorgen. Heating causes the insoluble residue to yield a wide range of hydrocarbons, aliphatic and aromatic, while amino acids in low concentrations (entry M2) have also been extracted from these meteorites.

Although carbonaceous meteorites are almost surely of cometary origin, and although microorganisms are extremely abundant inside comets according to our point of view, the meteorites do not at first sight seem promising objects in whch to search for evidence of extraterrestrial life. This is because meteorites are derived from the non-volatile fraction of comets. The relative abundances of H, C, N, O given in Table C1.1 are certainly extremely life-like, but refer to the volatile fraction, not to the unevaporated material from which the meteorites are derived. One would expect the latter to be largely the inorganic refractory component which remains after the life-like component has disappeared through evaporation. In order for them to reach the Earth, meteorites must have orbits with perihelion distances less than or equal to the radius of the Earth's orbit around the Sun. When near perihelion, meteorites (being small bodies) must be generally heated through their interiors by the Sun, not just once but millions of times. One would not think such a history favourable to the preservation of evidence of life-forms. Yet microfossils turn out to be remarkably indestructable, and already in the early 1930's there was a suggestion that fossilised spores were present in Orgeuil, a meteorite of Type I that fell in 1864. In the early 1960's this speculation became proposed seriously by G. Claus and B. Nagy, who claimed that certain 'organised elements' in Orgeuil and the Ivuna meteorite (type II) which fell in 1938 in Tanzania were of biological origin (G. Claus, B. Nagy, D. L. Europa, *Annals N. Y. Acad. Sci.*, 108, 1963, 580).

The claim of Claus and Nagy provoked instant uproar, with the critics braying so stridently that the world became convinced the claim must be wrong. Claus reneged, and Nagy retreated while continuing to hint in his writings that it might be so, rather in the style of Galileo's whispered " E pur si muove" This particular controversy passed over our heads at the time, and like the rest of the gang we came to believe consensus opinion, until 1977 when at last we had occasion to read what Claus and Nagy had actually written. The puzzle we saw immediately was that 'organised elements' had been found only in a handful of meteorites. If they were the result of contamination, spores in the wind, why had the contaminants elected to penetrate only a few meteorites and to leave the majority alone? Perhaps because the few were porous and the majority were hermetically sealed? But if you tried to use this same argument the opposite way around, against consensus opinion, just the same voices would bray equally loudly that no piece of stone is hermetically selaed against penetration by bacteria. As Mozart unwisely observed there is no limit in the world to the number of ass-ears, for which impertinence they buried him in a pauper's grave. In these later days of soi-disant 'peer-reviews' his fate would likely have been worse.

One of the present authors (N. C. W.) wished in 1977 to write a little in favour of Claus and Nagy, but the other saw no point in deliberately stirring-up a hornet's nest. However, on reading in detail everything to do with the controversy, F. H. decided the issue had been by no means as one-sided as he had been led to imagine. So he permitted the following passage to appear under joint authorship* :

> Strong arguments to support the view that these fossil structures are indigenous to carbonaceous chondrites were put forward by Klaus, Nagy and (later by) Harold Urey. The concentrations of fossils are remarkably high in all cases studied and account for at least ten percent of the insoluble organic matter . . . .Many questions, however, remain to be answered, the most immediate one being how such organisms could evolve and become fossilized within the fabric of these meteorites. In the mid-1960's the lack of a convincing answer to this question led many scientists to become sceptical about the fossil explanation of the so-called organised structures in meteorites.

> An alternative explanation was that these fossil-like structures are mineral grains which have acquired coatings of organic molecules by some non-biological process. The difficulty with this theory, though, is that the highly organised cell-like appearance of these structures would still remain a mystery. Terrestrial contaminants were suggested as another possibility, but this is unlikely to be the correct explanation for the majority of structures because of . . . (the reason already given above).

There the situation remained until it was taken up early in 1980 by H. D. Pflug. Within a month or two, Pflug had found a similar profusion of 'organised elements' in thin sections prepared from a sample of the Murchison meteorite, a carbonaceous chondrite of Type II which fell on 28 September 1969 about a hundred miles to the north of Melbourne, Australia. He presented his initial results at the annual out-of-town meeting of the Royal Astronomical Society, held in 1980 in Cardiff, leaving his audience to make of it what they would, expressing no opinion himself. What was needed he told the meeting was an improvement of technique, an improvement of the signal-to-noise ratio in the experimental method.

Lifecloud J. M. Dent. 1978, page 111.

9-

The method eventually adopted by Pflug was to dissolve-out the great bulk of the minerals present in a thin section of the meteorite, doing so in a way that permits the insoluble carbonaceous residue to settle on a piece of film, rather as glacier till accumulates in a moraine as the ice which has hitherto been holding it melts. The trick lies in being able to examine the resulting 'till' in an electron microscope *without disturbing it from outside*. The carbonaceous 'till' can also be examined with an ultraviolet probe, by a laser Roman probe, and by a mass spectrometer, all with a resolution spot of diameter ~ 1 μm. The mass spectrometer permits one to ensure that the few minerals which might have survived the use of strong hydrofluoric acid, chromite or spinel, are not present in a particular preparation, and therefore that a particular electron micrograph refers to the carbonaceous residue alone.

Using the probe techniques as supporting evidence, Pflug's main attack on the problem has been by morphological comparisons, the method which paleontologists mostly rely on. This method has the minor disadvantage that it demands a little prior knowledge if one is to evaluate the data meaningfully. One needs to know for instance that the object at the right of Figure M1.1, found in the Gunflint chert from N. Minnesota, is a fossilised microfungus in order to appreciate that the similar object at left, found in the Murchison meteorite, is also a fossilised microfungus. The sceptic, irreversibly programmed not to accept evidence such as this, argues that anything might be deduced from morphologial comparisons. Anything ? Let the sceptic try Figure M1.2. On the left is a strange *terrestrial* iron-oxidising (and manganese-oxidising) bacterium. At the right of the Figure M1.2 is a morphologically similar structure found in the Murchison meteorite. Is this a chance similarity of shape ? ' No ', even the most determied critic would probably answer, ' it is contamination '.

By storing a preparation in water for two or three weeks, contamination can be induced deliberately, and then the contaminant bacteria can be examined and compared under the electron microscope with objects such as the one at the right of Figure M1.2. When the focus of the microscope falls appropriately with respect to the surface of the object at the right of Figure M1.2, the surface is seen to be sharply defined. The surfaces of induced contaminants are soft, however, because of the thin layer of slime with which they are coated. Besides which, the scale of the meteoritic object is about three times smaller than the terrestrial microorganism. Futhermore, a considerable fraction of the carbonaceous residue in the meteorite shows biological characteristics. For example, the structure of Figure M1.2 is identical in detailed shape to a recent microorganism, pedomicrobium. This structure is evidently of similar origin to the one at left in Figure M1.3, which is an outlier of the cluster shown in the upper part of Figure M1.3. The sheer profusion of biological forms in the Muchison meteorite precludes them being contaminants.

The sceptic's views are in any case irrelevant, since, with the technique now established, data like that shown in Figure M1.1, M1.2 and M1.3 can be piled-up almost without limit, until it becomes overwhelming. Indeed, if one looks through the full range of the data obtained by Pflug, it is overwhelming already.

Fig. M 1.1. Microfossils in the Archean sediments of the Earth compared with meteoritic micro-fossils (courtesy Prof. H. Pflug)

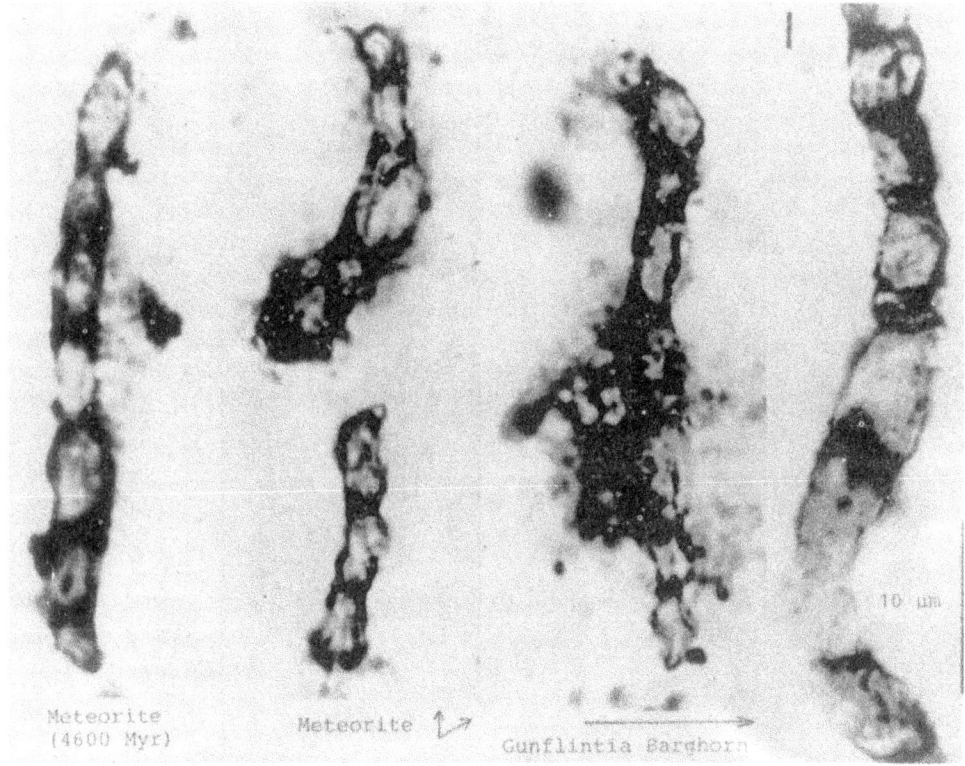

Fig. M1.2. Microfossils of iron oxidising microorganism in the Murchison meteorite compared with similar present-day organism.

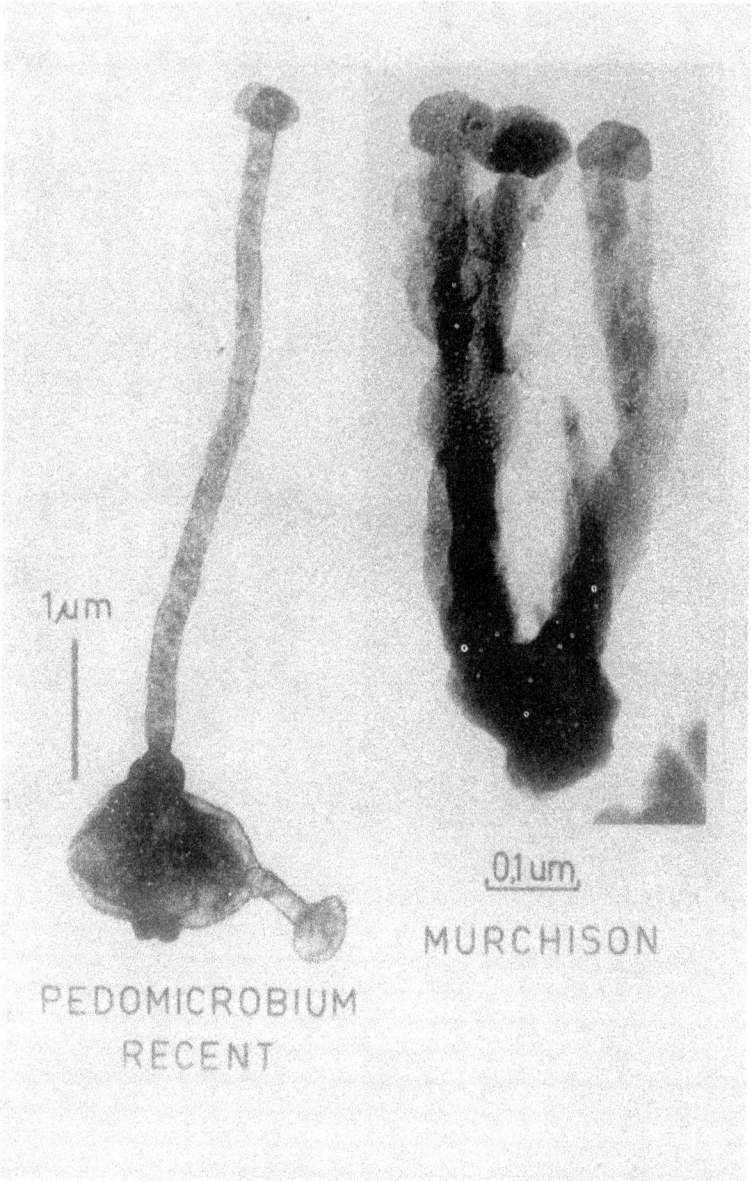

PEDOMICROBIUM
RECENT

MURCHISON

Fig. M 1.3. Clusters of iron oxidising microorganism fossils in the Murchison meteorite.

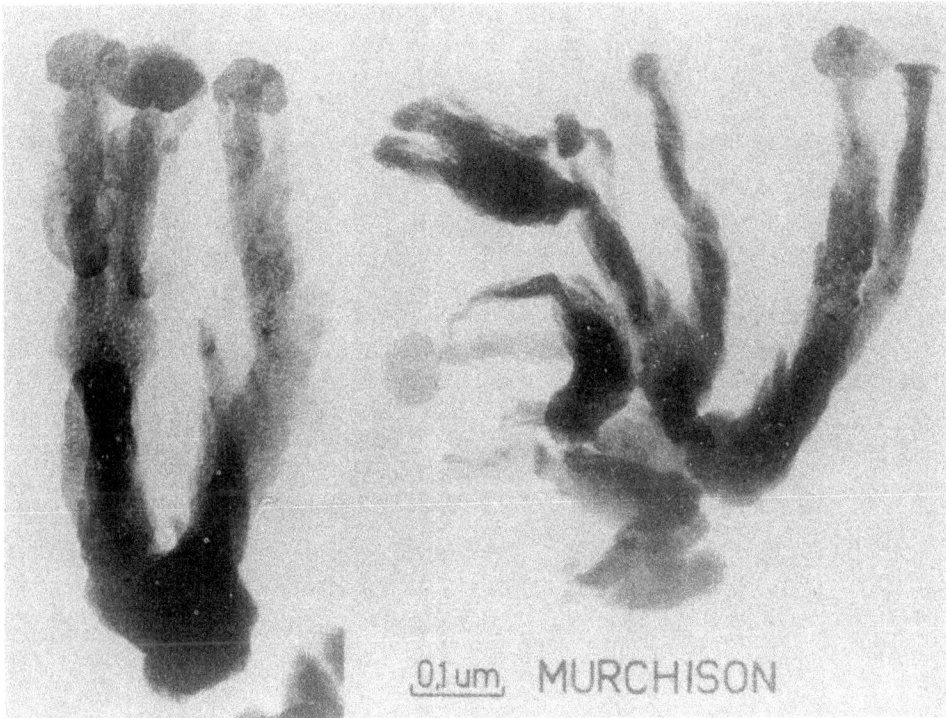

0.1 um MURCHISON

M 2.   The Stereochemistry of Amino Acids in Meteorites

The stereoisomerism of amino acids in meteorites is a tangled story. Even the starting point is not as clear-cut as it is usually said to be. According to the coffee-cup science practised by government funding agencies, biological proteins are made up from a subset of 20 amino acids of the left-handed levorotary form (L). If one keeps to the proteins actively used in the operation of biological cells this statement is nearly true (like saying 10^6 is a negative number) but it is not exactly true, since in one case known to us – a protein in the slime trail of a snail – the amino acids are dextrorotary (D), and there may be other cases as well. More importantly, D–amino acids appear widely in the inactive structures of cells, particularly in the mucopeptides of cell walls, where D–ala and D–glu are about as common as the levo forms of these amino acids. The D-form of glycine also appears. Moreover, amino acids other than the subset of 20 occur in the mucopeptides, M–DAP, LL–DAP, L–Orn and L–hser are examples.

All this is *in vivo*. During the fossilisation of biochemical materials, amino acids take part in complex chains of chemical reactions in which so-called non-biological amino acids (outside the subset of 20) are produced. Also during fossilisation there are switches of stereoisomerisms so that L–forms mostly become a mixture of L and D, tending asymptotically towards an equal racemic mixture. These processes have been studied with particular regard for the degradation of biomaterials under terrestrial conditions (P. E. Hare, " Geochemistry of Proteins, Peptides and Amino Acides ", in *Organic Geochemistry*, ed. S. Eglington and M. T. J. Murphy, Longman, 1969). Similar degenerative processes must occur in meteorites, with greater complications one would think on account of the millions of times meteorites are heated and cooled.

Concentrations of a few parts per million by mass of the more abundant biological amino acids have been found in carbonaceous meteorities. Early reports were of nearly equal D and L forms in the Orgeuil and Murchison Meteorites (J. G. Lawless, K. A. Kvenvolden, E. Peterson, C. Ponnamperuma and E. Jarosewich, *Nature*, 236, 1972, for Orgeuil ; and for Murchison, J. G. Lawless, *Geochimica et Cosmochimica Acta*, 37, 1973, 2207). A similar report of equal D and L forms has given more recently for a carbonaceous meteorite recovered from the region of the Yamato Mountains of Antarctica (A. Shimoyama and C. Ponnamperuma, *Nature*, 282, 1979, 394). It has been argued that since terrestrial contamination, if it had occured, would have yielded a preponderance of L–forms, the essentially racemic mixtures in Orgeuil and Murchison must be of extraterrestrial origin. The argument has been pressed further, incorrectly as we have always thought, to the point of saying that racemic mixtures show the amino acids in question to be of non-biological origin. This is a *non-sequitur* because of the possibility of racemerization occurring during the fossilisation of microorganisms. And because of the complex history of meteorites it seems to us hard to predict exactly what would happen to amino acids within them, particularly as neither laboratory data nor paleontological evidence covers the physical conditions adequately.

126

However the argument on this issue may now be moot. Very recent determinations for Murchison by M. E. Engel and B. Nagy *(Nature, 296, 1982, 837)* **are** set out in Table M2.1 .

TABLE M2.1
Murchison Meteorite Amino Acid D/L Values

Extract	Glu	Asp	Pro	Leu	Ala
$H_2O^{(1}$	0.322	0.202	0.342	0.166	0.682
$H_2O^{(2}$	(±0.044)	(±0.005)	(±0.065)	(±0.021)	(±0.062)
$HCl^{(1}$	0.30	0.30	0.30	ND	0.60
	(±0.02)	(±0.04)	(±0.02)		(±0.03)
	0.176	0.126	0.105	0.029	0.307
	(±0.013)	(±0.004)	(±0.017)	(±0.002)	(±0.010)

Two different methods were used to obtain these values (1 and 2), for the details of which see the reference cited. The HCl extract was obtained from material which had been subject to water-extraction, and so the HCl extract refers to amino acids more strongly bound in the meteorite than the amino acids which came out already in the water extract.

In 1971, J. Oro, S. Nakaparksin, H. Lichtenstein and E. Gil-Av *(Nature, 230, 1971, 107)* had given 0.29 for the D/L ratio of alanine in the Orgeuil meteorite, far from being racemic as claimed by **Lawless** *et al* in 1972. Oro and his colleagues sought to explain their result in terms of contamination. They argued that a racemic mixture of alanine in the meteorite itself had become mixed with L-alanine from microorganisms of terrestrial origin which had invaded the meteorite over the century or so since its fall in 1864. This explanation of 0.29 for the D/L ratio would require a close coincidence between the amounts of alanine from terrestrial and extraterrestrial sources, a coincidence which Oro *et al* preferred to accept rather than face the alternative discussed below.

Engel and Nagy dispose of the ambiquity of contamination in an elegant and apparently decisive way. They find none of the four biological amino acids Thr, Tyr, Phe, Met, which should be present if there had been significant contamination by biological proteins of terrestrial origin. A very slight contamination was indicated by a barely detectable quantity of Ser, the major amino acid found in fingerprints, which Engel and Nagy attribute to the handling of Murchison, before a sample of it reached their laboratory ! The results of Engel and Nagy are readily explained in terms of the degenerative processes which produced the carbonised microfossils of entry M1, degenerative processes which partially racemerized the amino acids of Table M2.1, and which destroyed the Ser, Thr, Tyr, Phe and Met. Engel and Nagy offer as an alternative, the thought that perhaps amino acids of extra-terrestrial but non-biological origin, amino acids that were initially racemic, had experienced a selective destruction of their D-forms. In making this proposal it seems that Nagy was trying to make amends for being correct in his pioneering work of the 1960's (entry M1).

127

O

Origin of life

O 1. Precambrian Life

The appearance in profusion of multicellular life, the metazoans, in the fossil record, was such a marked phenomenon that geologists chose long ago to adopt this dramatic bilological development as the definition of the end of the Precambrian era. Why it took so long in the Earth's history (some 3800 million years) before the metazoans appeared has for long been a matter of speculation. The physicist looks naturally for a physical explanation. Starting with the likely **astrophysical** point that the luminosity of the Sun during the early history of the Earth was some 25 per cent *less* than it is at present, together with the geological point that the Earth was nevertheless not permanently glaciated, one infers that the atmosphere of the Earth must in the Archean and perhaps also in the Proterozoic have generated a very strong greenhouse effect. A high atmospheric CO_2 pressure seems the most likely cause for such an effect, and indeed if all the CO_2 present in the **carbonate** rocks of the Earth's crust were put into an early atmosphere, the resulting greenhouse effect would have been strong enough to push the mean terrestrial temperature above 50°C, the lower luminosity of the Sun notwithstanding. One then notices that the usually postulated time sequence of lifeforms—procaryotes, single-celled eucaryotes, and eventually multi-celled eucaryotes—is a sequence of decreasing tolerance with respect to temperature. From this line of reasoning it seems attractive to argue that the more complex forms of biology had to await a lowering of the Earth's mean temperature, which took place as the atmospheric CO_2 gradually became fixed into carbonate rocks.

Unfortunately for this tidy physical view of things, there were glaciations in the early Proterozoic, some 1,500 million years before the emergence of the metazoans. Perhaps there was first a fixing of all the atmospheric CO_2, followed by a considerable return of the CO_2 back to the atmosphere. Perhaps indeed there was a sequence of wild swings of the atmospheric CO_2 between higher pressures and low pressures, giving glaciations during the low periods and preventing the evolution of complex biological forms during the high periods when temperatures were lifted by the atmospheric greenhouse effect to ~ 50°C.

A biological view of things suggests something very different. The metazons are usually said to have evolved from the protozoa, but protozoa do not reproduce themselves in a fashion to suit a metazoan. To grow an ordered properly anchored surface layer of cells (a **necessary** property for even the humble jellyfish) it is important geometrically for the process of cell division to generate four daughter cells from the parent cell. Neither budding nor binary fission (as in bacteria, microfungi and many protozoa) will do. Some protozoa have more than two cell nuclei, however, with each nucleus passing to a daughter cell on replication. This comes nearer to what is required, but it still does not satisfy the need to generate four, and only four, daughter cells. Perhaps it was the development of this critical property which at last set the stage for the emergence of the metazoans.

So one might come to a tidy biological view of things, according to which the Archean was the era of the procaryotes, in which eucaryotes including the protozoa evolved from the procaryotes during the upper Proterozoic, and in which the trick of fourfold cell division appeared in some protozoan towards the end of the Precambrian, a protozoan from which the metazoans then evolved. Unfortunately for this scheme, the time intervals between the major steps are far too long, unless indeed there was a need for the physical environment to change, as with the declining temperature sequence **considered above.** Biology inherited an illusion from Charles Darwin, namely that long **time-scales** were mysteriously helpful to evolution. Except in the physical sense just mentioned, this is untrue. The sheer complexity of biochemical systems forces just the opposite conclusion, as will be shown in entry O3. If an exceedingly conmplex system can evolve by internal processes in a long time, 1000 million years, a system of only marginally smaller complexity can evolve in a trivially short time.

This conclusion turns out to be supported by the paleontological facts which vitiate the tidy biological view. **Figures O1.1 and O1.2** show eucaryotic microfossils from the Swartkoppie of S. Africa and from the Isua series of SW Greenland, not in the

130

Figure O 1.1.–Eucaryotic micro-fossils from the Swartkoppie sediment of S. Africa. Data by courtesy of Prof. Hans Pflug.

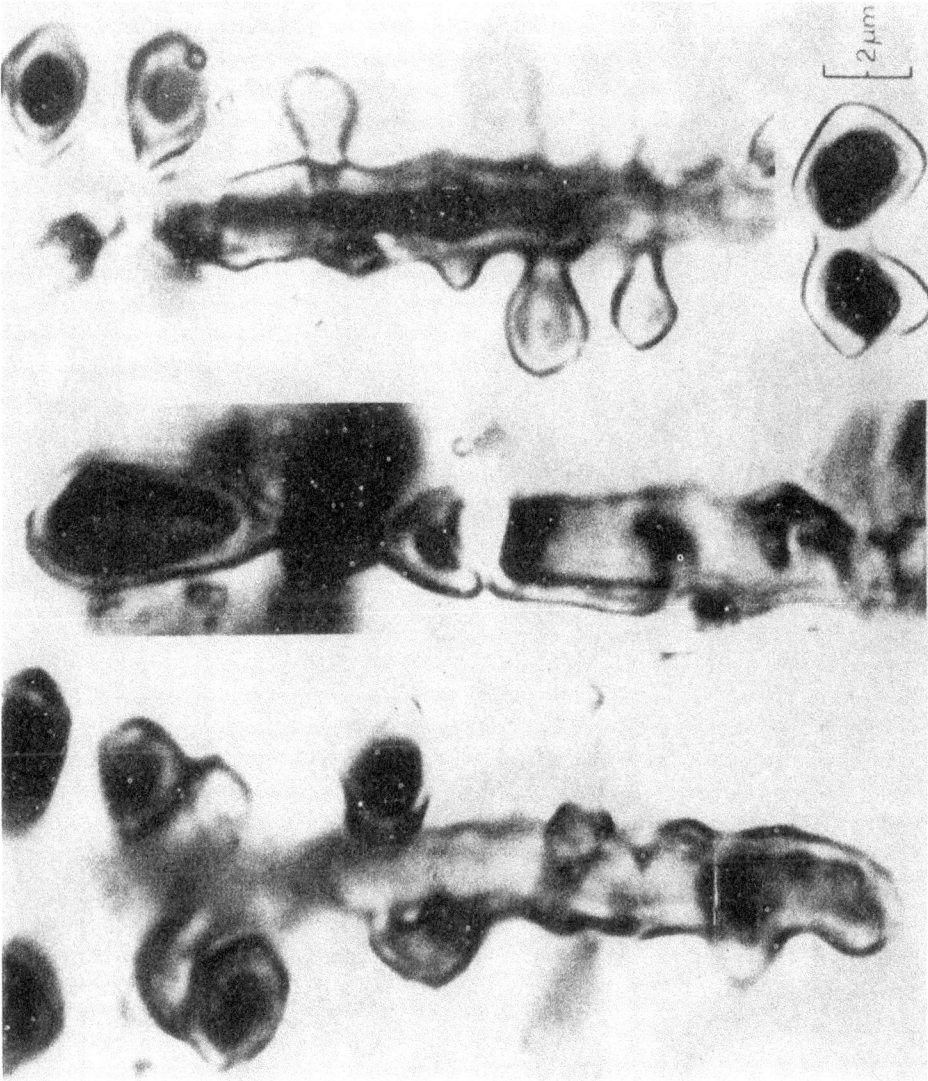

131

Figure O 1.2.–Eucaryotic microfossils in Archean sediments. Data by courtesy of Prof. Hans Pflug.

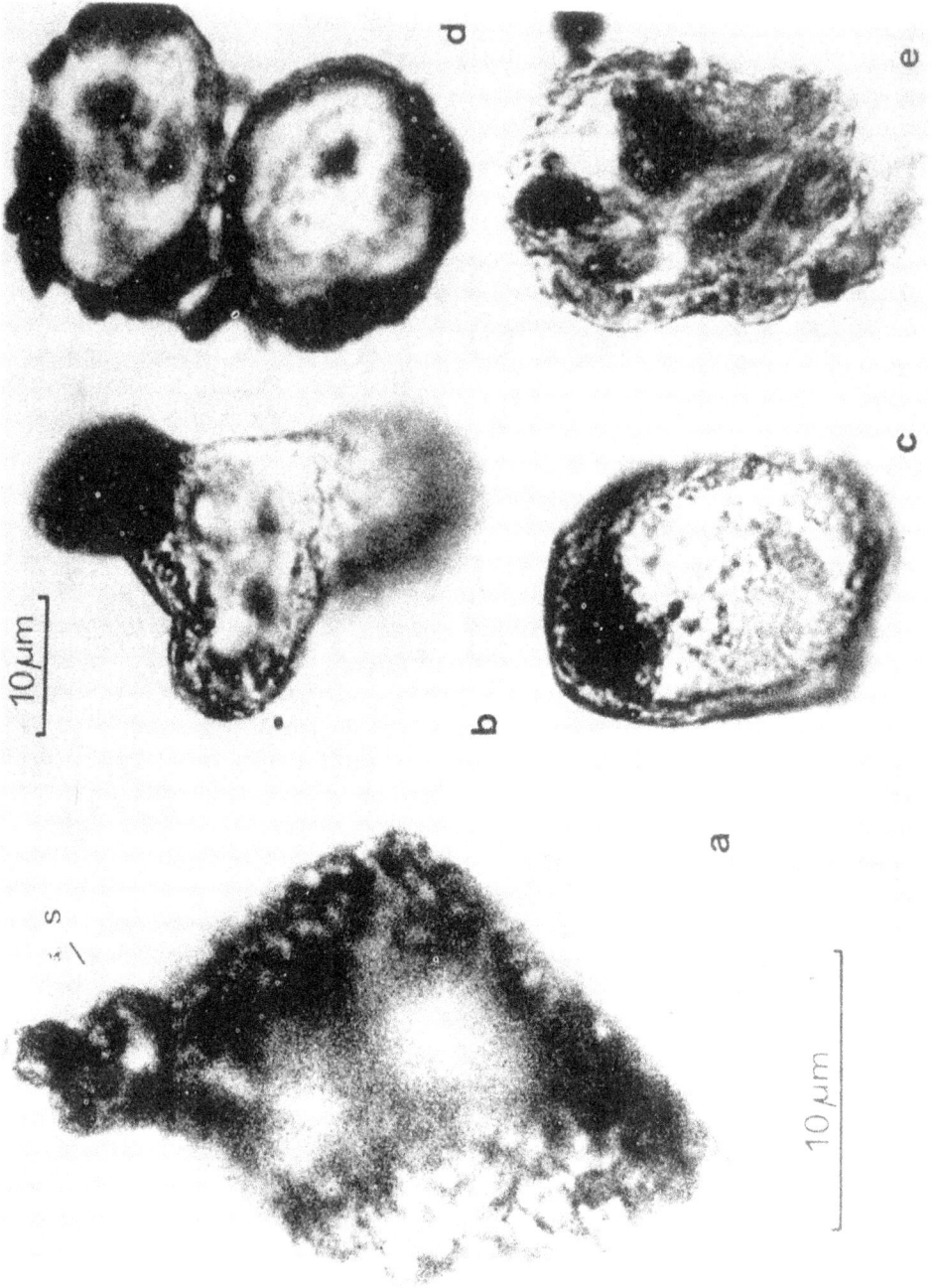

Proterozoic but as far back into the Archean as geology goes (H. D. Pflug and H. Jaeschke-Boyer, *Nature,* 280, 1979, 483). It is interesting to consider the ciriticms of these results which have been put forward, especially the results concerning the Isua series. The rock from which the specimens were obtained was said by some to be sheared, which it isn't, **as can readily be** proved from the crystal structures within it. The rock was said by others who had never seen it to be weathered, which it isn't. The rock has experienced medium-grade metamorphism, probably at a temperature of about 400°C. This would destroy genuine microfossils it was said by the Delphic Oracle, so that the objects of Figure O1.2 had to be of post-metamorphic origin. But they are not, as can be seen from the actinolite needle (a product of the metamorphic process) which penetrates the microfossil of Figure O1.3. Besides which, the infrared spectrum of *E. coli* given in Figure I 6.1 was obtained for bacteria in KBr disk with a temperature of 350° C at its centre and about 400° C at its circumference. Figure I 6.1 shows no degeneration at all of the characteristic pattern at ∼ 3.4 μm, a feature that is still recognisable (for specimens sealed inside an inorganic matrix) up to about 500° C.

The objects present in the Isua series were first examined by B. Nagy, J. E. Zumberge and L. A. Nagy *(Proc. Nat. Acad. Sci Washington,* 72, 1975, 1206) who identified them correctly as carbonaceous skins, who thought a biological origin possible but who preferred an abiologic origin (doubtless because of the uproar of the 1960's described in entry M 1). D. Bridgwater, J. H. Allaart, J. W. Schopf, C. Klein, M. R. Walter, E. S. Barghoorn, P. Strother, A. H. Knoll, and B. E. Gorman (*Nature,* 289, 1981, 51) identified the objects as fluid inclusions, thereby supporting the view that the more distinguished authors there are to a paper the sillier its contents. E. Roedder (*Nature,* 293, 1981, 459) thought the object to be derived from carbonate grains, probably of non-biological origin, a view which might have applied to the objects considered one-at-a-time in isolation from the rest of the evidence, evidence of the kind shown in figure M 1.1, the objects at right in this figure (from the Gunflint chert of N. Minnesota) being considered by experienced paleontologists to be an unquestioned example of a fossilised microfungus.

The objects in question are carbonaceous skins often encrusted with **carbonates** and they are indigenous to the rock (Figure -O 1.3). Pkflug considers them to be of biologic origin because they satisfy the following four criteria :

(1) The object has the size and shape of a cell, it consists of carbonaceous material, is often externally encrusted with carbonates and usually preserved in a silicious condition. The structure can be separated from the rock by HF-treatment and then remains as a coherent spherical or filamentous skin.

(2) The object occurs associated with other objects of the same kind and together with non-structured carbonaceous debris. The assemblages are preferentially arranged along bedding planes or other primary patterns of the sediment, which is usually a chert, shale, stromatolite or related rock.

Fig. O 1.3.–Isua microfossil showing actinolite needle going through it.

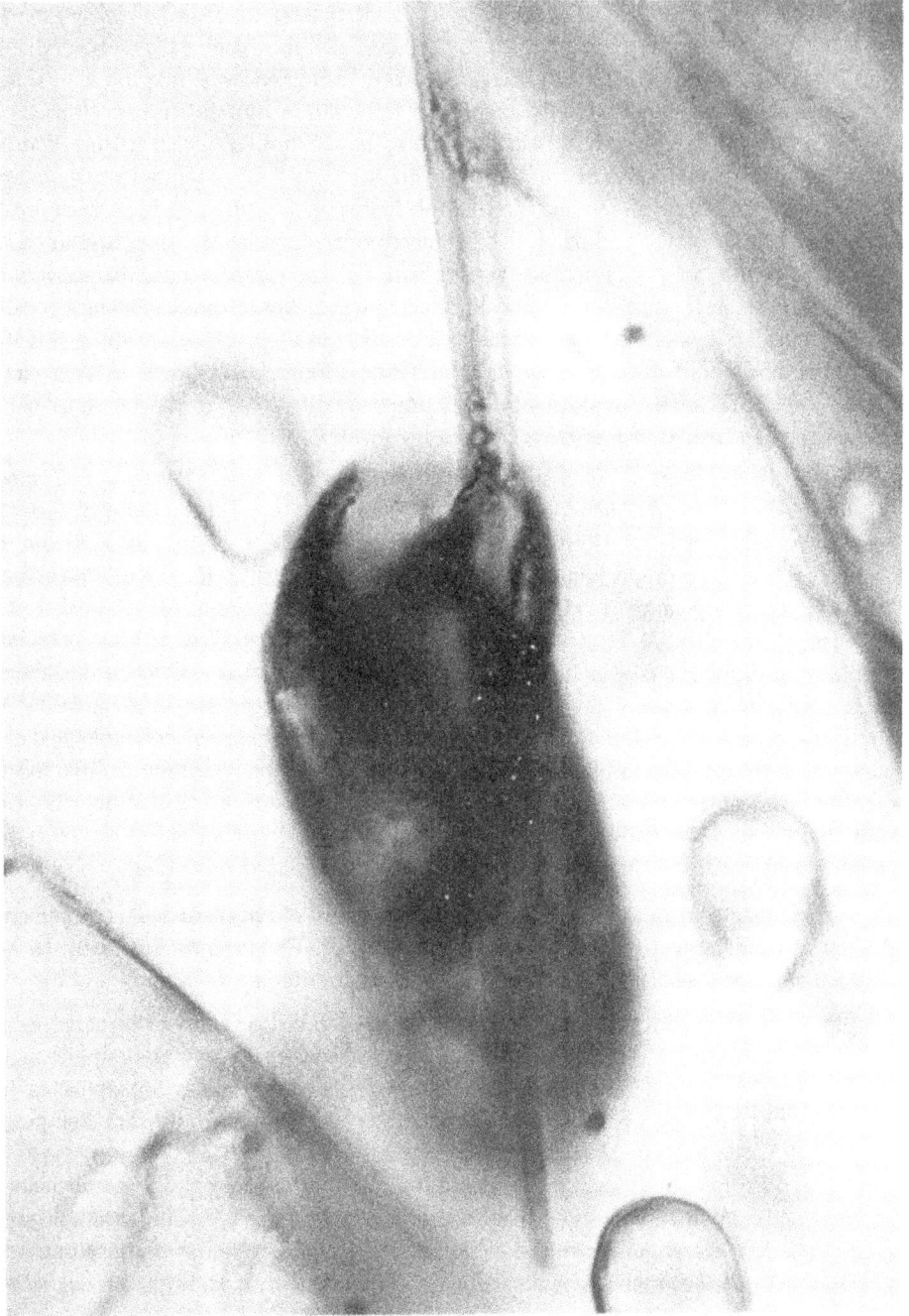

(3) Well-preserved specimens show structural details in the cell-wall or sheath (Figures M 1.1, O 1.1 and O 1.2). Others are present in typical stages of reproduction and growth. Individuals are often gathered together in colonies of definable interior arrangements.

(4) The host rock contains carbon which in its isotopic composition resembles that of biological material, and the host rock contains fossil organic compounds (kerogen etc).

There is little doubt that if the scientific world had been educated to believe that life is external to the Earth, to believe that life, whether procaryotic or eucaryotic, established itself here at the earliest moment it could be supported, people would find the above evidence entirely convincing. Anybody who doubted it would be thought defective in both reason and eyesight. Criticism has come because people have been miseducated to believe the opposite, to a point where beliefs have come to seem more important than facts. Yet the truth will out. In entry E9 it was mentioned that studies of the base-sequencing in the rhibosomal RNA of procaryotes and eucaryotes has failed to reveal the hypothetical evolutionary connection between them (e.g. Fox *et al.*, *Science*, 209, 457, 1980). Those who believe in the prejudices instilled by the educational system are turning out wrong. Those who simply believe the evidence of their eyes are turning out right.

O 2.   The Origin of Life, an Elusive Concept

Every generation of every species of plant and animal is preceded by a similar generation, like begets like as one says, an everyday observation for macroscopic species which Louis Pasteur extended to microorganisms. In view of this apparently inviolable law of succession the concept of evolution, of a branching whereby a species bifurcates into two or more different species, was a daring step. In the eighteenth century when the step was taken (entry E 1) nobody had every seen such a branching, and still today nobody has actually seen new species originate. In the fossil record species appear abruptly, out of nowhere it seems (entry E 3).

Beliefs for which explicit evidence is lacking turn out badly more often than not, even in cases where a belief happens to be correct, because it can wreck serious harm on the way. The trouble is that people feel impelled to invent a supposed reason, a trumped-up theory, to support an emotionally-held idea, and they are apt to seize on the first theory to come along if it happens to have a superficial air of plausibility. The theory must be right, the people say, because the idea it supports is true, and so a false theory comes to be believed with some passion. When facts eventually go against the theory, the facts are put out of mind instead of the theory, and the thing eventually becomes an immense intellectual burden, preventing meaningful progress.

135

It would clearly have been better in retrospect if those who became convinced of evolution in the eighteenth and early nineteenth cunturies had been prepared to keep an open mind as to its cause. The tautology of natural selection, the Alladin's lamp of the late nineteenth century, would then have been avoided (entry E1):

If among a number of varieties one is better fitted than the others to survive in the environment as it happens to be, then it will be the variety that is best fitted to survive that will best survive.

It should have been apparent to classical biologists that the great diversity of biological forms could not flow from a trivium of this minor order. Instead, the trivium was elevated to the status of a god before whom one was obliged to bow down and worship, a procedure that did nobody's wits any good. It should have been clear from the failure of the fossil record to reveal the secret of evolution that something unexpected had to be involved. The way would then have been open for modern microbiology to show (from the complexities it soon revealed) that the issue was cosmic at root, likely enough more profound than anything yet known to the professional cosmologist.

Unlike classical biology, modern microbiology has quantifiable aspects, not in its full measure of complexity to be sure, but in the relatively simple issue of the orderings of amino acids in a particular subset of polypeptides, the enzymes, chosen out of the full set of polypeptides for the following reason.

Code words are used for many purposes, in war, as the key to a safe-deposit, a calling-instruction in a computer program, and as a so-called intron for the expression of a gene. It is the nature of a code word that one word would in principle serve as well as another, yet by an arbitrary choice we elevate a particular word to special significance. So it may be with a considerable number of polypeptides. At least 100,000 with well-defined amino-acid sequences are used in the operation of the brain, but because of the possibility that they may mostly be code words, quantitative significance cannot be inferred for this very large subset. There are others, however, that are certainly not code words, in particular the enzymes which interface a biological system to its environment.

In other entries we have come on many examples of enzymes which interface the environment, those responsible for repairing damage due to X-rays and ultraviolet light (entries B2 and B3), those responsible for the respiration of sugars (entry E9), those used in chemoautotrophic bacteria for catalysing a wide variety of chemical reactions (entry B6), those used to break linkages in **polysaccharides**, not the same for every polysaccharide, starch for example being considerably different from cellulose—and so on over a wide spectrum of activities.

In this connection it may be noted that sugars can be thought of as a number of formaldehyde groups, $(H_2CO)_n$, with exchanges of atoms between the groups occurring as they are linked together into the arrangements which distinguish individual sugars. Since hydrogen, H$_2$, and carbon monoxide, CO, are the commonest of molecules, the ultimate ingredients of sugars are the dominant chemical substances in the universe. Their existence is not an internal invention of biology. Biology therefore had to discover upwards of some fifty externally-defined enzymes in order to realise the potential of sugars as an energy source.

Although some enzymes are employed internally within cells rather than at the environmental boundary, enzymes operating internally are frequently engaged in processes that are fundamental to carbon chemistry itself, and so are not arbitrary. The enzymes that unzip polypeptide linkages are examples. (While one might conceive of polymeric linkages different from the peptide bond being used as an alternative to existing biology, the number of such alternatives cannot be enromously large within a chemical system based largely on the properties of only four elements, H, C, N and O.) Plainly then, enzymes are not arbitrary code words.

The main problem for those who believe life to have arisen spontaneously in some brew of non-biologic substances is not the origin of the amino acids themselves. Nor is it simply to generate polymerization among amino acids. The problem lies in the specificity of the chains which define the enzymes, and of bringing the many chains necessary for the operation of a biological system into conjunction with each other. How specific, let us ask, must a chain of amino acids be in order that it can serve as an enzymic catalyst in some chemical reaction, as for instance in the situation shown schematically in Figure O2.1 ?

To function as an enzyme the globular protein into which the polypeptide coils itself has to possess a site at its surface where the external chemical substances can attach themselves in suitable juxtaposition, and also where the detailed nature of the site sets up intermediate reactions between the external substances and atoms of the enzyme itself. At the end of these reactions the situation must be such that the substrate has been transformed and the atoms of the enzyme have returned to their intitial arrangement. For quite complex substances forming the substrate in most cases, these are not simple requirements, as is clearly proved by the ease with which a biological mutation involving a change of only a single amino acid can destroy the function of an enzyme. This happens when a mutation hits an amino acid which affects the operation of the active site illustrated by the example in Figure O2.1. The variability across the whole face of multicellular biology of the enzyme **cytochrome–c** suggests that the chance of a chain of 104 amino acids (chosen randomly from the restricted set of twenty ' biological ' amino acids) happening to yield a viable enzyme with the properties of cytochrome–c is about one part in 10^{80} (entry E9)

Such a minute chance applied to only one crucial enzyme is sufficient in itself to rule-out all possibility of a biological system arising from naturalistic processes within a so-called primordial organic soup. Those who are impelled at all costs to defend the organic-soup theory are therefore obliged to reduce very greatly the chance of arriving at

137

10 –

each of the many essential enzymes, to as small a number as their consciences will bear. Pretty well the ultimate in elasticity of consciences will bear Pretty well the **ultimate** in elasticity of conscience is to be found in H. Quastler, *The Emergence of Biological Organization.* Yale University Press, 1964. Quastler considers that it might be sufficient to have merely two particular amino acids in juxtaposition with each other, with a chance of one part in 400 of obtaining a surface site like **that of Figure O 2.1,**This opinion was given short shrift by F. B. Salisbury (*Nature.* 224, 1969, 342) who pointed out that the appropriate juxtaposition for every enzyme would then be found to occur on every polypeptide of appreciable length, an incorrect deduction since the great majority of polypeptides have no observed enzymic function at all. Salisbury considered that 15 amino acids must minimally be correctly ordered to yield an appropriate surface site like that of **Figure O 2.1.** in which case the chance of obtaining a particular enzyme in a random protein chain of moderate length would be of the order of one part in $(20)^{15}$ Since

Fig. O 2.1.–Enzyme action.

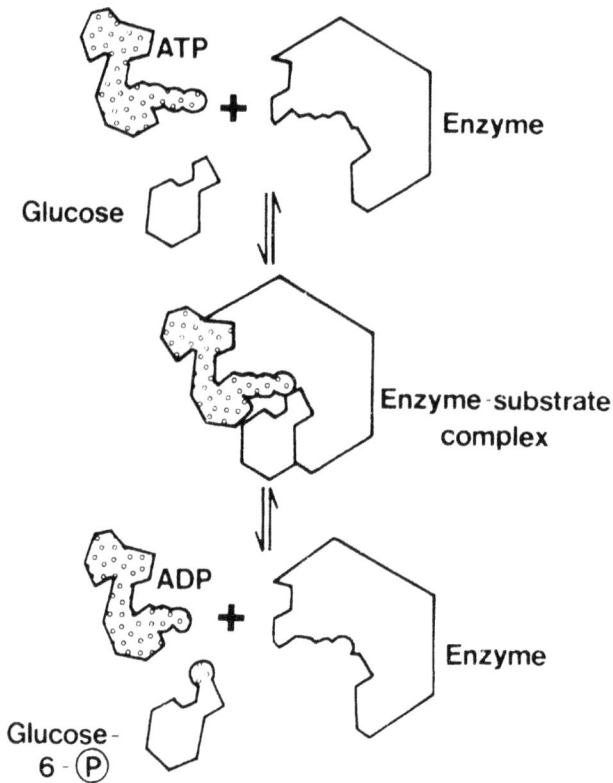

138

this is not nearly as small as the chance of one part in 10^{80} which the observed variability of cytochrome–c implies, a requirement for the positioning of as few as 15 amino acids should be considered a pretty fair elasticity of conscience. Indeed Salisbury remarks :

> . . . we have determined the structures of a few enzymes, and it does *not* appear that the sequence numbers for an active site are small. In lysozyme, for example, the substrate binds to the enzyme through at least six hydrogen bonds and more than 40 somewhat weaker contacts ".

One way to represent the information content of life is by the ratio of the number of nonsense arrangements of polypeptides to the number of possible living arrangements. If we keep only to the **subset** of the enzymes, and if we take this ratio for a single enzyme to be no more than 10^{20}, i.e. approximately $(20)^{15}$, the combined ratio for the whole subset of 2000 enzymes is $(10^{20})^{2000}$, a number with 40,000 digits, a number so vast that it would occupy ten to twenty pages of print if written out in longhand form, 1,000,000, . . . The enigma of this huge number bestraddles the problem of the origin of life.

O 3. Stirring the Soup

Let us make a number of assumptions favourable to the idea that life originated here on the Earth, starting from a broth of organic materials of abiological origin. Let the soup have a volume as great as the whole world ocean and let it have high concentrations of amino acids in particular. Let the amino acids polymerize in some way into chains of whatever lengths one pleases. It makes no difference to the following argument whether the amino acids join together on their own account or whether one supposes a primitive templet system directs their polymerization, since the problem to be considered is one of ordering in the polymers not the process whereby they are formed.

There is a limit to the number of concessions we are willing to make, however. Having once joined together, the amino acids do not separate in the initial absence of an enzymic system, because of course the peptide linkages are strongly bonded. Intense ultraviolet light might break the linkages, but only at the expense of destroying all attempts to secure a delicate biological development in the broth. Once formed, the structures of the polypeptides are fixed, they do not twinkle as some authors assume them to do. For a whole world ocean of broth, and for polypeptides averaging a hundred amino acids, say, there would be $\sim 10^{43}$ chains, after allowing for the excess of water that is necessary if the polypeptides are to have mobility.

Some of the polypeptides so formed may be capable of enzymic action, one in $\sim 10^{20}$ according to the stretching of our conciences achieved in entry O2. On the latter basis the problem is to get the enzymes together, since on a molecular scale they are widely separated. Except in rare cases each potential enzyme is separated from the next one by $^\sim 10^{20}$ non-enzymic polypeptides. In rare cases, two potential enzymes will happen to be in **juxtaposition,** but there is unlikely to be any case in which three are in juxtaposition, unless one stretches one's conscience further, so that potential enzymes are more frequent, say a fraction 10^{10} instead of 10^{20} of the polypeptides, when as many as four or five of them could be in juxtaposition in the rarest case, but nothing like the fifty enzymes responsible for the respiration of sugars in actual biological cells.

At this stage it is necessary to set up a hunting process in which potential enzymes have a crucial property not possessed by others of the polypeptides. They recognize each other and stick together. An initially small aggregate of enzymes, two, three, four or five of them according to one's postulates, stick together as they move with respect to the huge majority of non-active polypeptides in their neighbourhood. The motion continues until it happens that a further potential enzyme comes into **juxtaposition,** whereon the further enzyme is recognised and added to the aggregate. And so forth for further potential enzymes until ultimately the aggregate grows large enough to permit an early stage of biological activity.

Some comentators have worried about the time involved in such an aggregation process, and have sought to improve the position in this respect by having the aggregate be reproduced a large number of times, which of course gives more chances for mopping-up potential enzymes. But as we see the situation, this is beside the point. The essential feature of the hunting process described in the previous paragraph is that it is, in effect, a description of how we ourselves would go about collecting-up a package of needles which had become scattered **throughout a haystack.** We would sort through the haystack, using our eyes and brains to tell us which was a length of straw and which a needle. One might make a machine to do the same thing, pulling-out the rare needles using a magnet—if they were of steel and not copper. If the needles were non-magnetic, however, a more complicated machine would be needed, and if the needles were made of some strawlike material differing only slightly from actual straw the process of separation would become pretty well impossible for a machine, and even quite difficult for the human brain and eye.

The real issue is that the hunting process described above is a piece of semantic trickery which conceals the critical point that intelligence is really being postulated, intelligence which distinguishes a tiny fraction of the polypeptides from the rest. But could not the behaviour of chemical systems, in particular the behaviour of polypeptides, simulate the operation of an intelligence ? It does so in our own brains one might argue, so why not at the more primitive level of an organic soup ? To answer these questions precisely it would be necessary to solve the **Schroedinger** equation for a many-particle system in fantastic detail, and this unfortunately we cannot do. But at least the situation is fair and squarely out in the open. To give the organic soup theory any chance of working, any chance of yielding the ordering of the enzymes, of coping with the number having 40,000 digits arrived at in entry O2, it is necessary to suppose the behaviour of chemical systems at the level of polypeptides to be capable of simulating an intelligence. Simple aggregates of very primitive enzymes have to be capable of 'recognising' other primitive enzymes which they have not 'seen' before, 'recognising' in the sense of distinguishing them from an enormous number, $\sim 10^{20}$, of other polypeptides. Against a person who claims it is so, there is nothing further to be said in argument. To proceed further we must appeal to experiment.

If there were some deep principle of nature which drove organic systems toward living systems the existence of the principle should easily be detectable in the laboratory. This applies whether the principle is one of 'seeing' and 'recognition' in the sense described above, or of concealed intelligence in other forms. One might seek for instance to claim that when amino acids polymerize into chains their orderings are not random, and likely enough it is true that the orderings are not completely random. But orderings that are not completely random remain a far cry from supposing that amino acids 'know' how to link themselves together so as to produce the enzymes and other critical polypeptides Such a notion of self-instruction by amino acids is an obviously wild proposal, but to disprove it decisively one must again turn to experiment. The ratio of the volume of the whole ocean to a chemist's test-tube is a number with only some 22 digits, so that using a test-tube of organic soup instead of the whole ocean of organic soup postulated in conventional biology, should merely lop 22 digits off the 40,000 digits which represent the information content of the enzymes, leaving 39,978 digits, essentially the same number as before. Nor does the length of time of an experiment matter significantly, even if the process of the origin of life were very strongly accelerating, say like the hundredth power of the time, $(time)^{100}$ Thus the reduction in the information accumulated in an hour instead of 1,000 million years would then be a number with some 1,300 digits, which would merely reduce the original 40,000 digits to 38,700, an information content that should be overwhelmingly detectable. An experiment done in half-a-morning, starting from simple organic ingredients, should therefore generate most, if not all, of the explicit structures of the enzymes. Needless to say, no such experiment has been successfully performed, showing that the enzymes did not arise by self-instruction of self-recognition, if indeed a disproof of these rather absurd ideas were needed.

Of all the facts available to us, whether in biology, chemistry, physics or astronomy, the huge information content of living system must surely be the most important, just because its numerical representation is so much larger than any other quantity with which we are familiar. A count of all the atoms in all the galaxies visible in the largest telescopes only yields a number with some 80 digits, which is less than the number of wrong ways of making even a single quite short-chain polypeptide such as histone–4. Thus if one were allowed a random trial of amino acid arrangements for every atom in the universe one would still be most unlikely to discover histone–4. Perhaps other

141

polypeptides might have served the function of histone–4 equally well one might argue, but if so mutations have never found them, since histone–4 seems to be essentially unique across the whole face of biology.

O 4.　The Virtue of Uncertainty

Unlike the other entries in this preprint, the present one is not directed at an explicit topic, nor does it attempt to be a self-contained argument. If the conventional theory of the origin and evolution of life on the Earth really fitted the facts in a precise way, it would make some sense trying to escape the huge number, $10^{40,000}$, estimated in entry O2 for the information content of life, perhaps along the lines considered in entry O3. But since the situation is quite otherwise (as we have shown repeatedly by many distinct arguments) it makes sense to take the huge number at its face value and to consider what the implications might be. The immediate **conclusion** that life must be deep-rooted in the **universe** is supported by the biological properties of interstellar grains (entries I) and by the existence of similar grains in other galaxies, outward from our own galaxy to the greatest distances that the largest telescopes can penetrate. One has to suspect that life may play a major role in astronomical processes which have hitherto been thought of as entirely mechanistic, a role perhaps reaching even into the fundamental aspects of cosmology.

We have sometimes been asked if anything is gained by converting the origin and evolution of life from a neatly-confined **terrestrial** problem to a still largely undefined cosmological problem, as if to suggest that our point of view is a counsel of despair. If there were equal chances of arriving at correct conclusions either way then of course one should investigate the simpler possibility first, but the simpler possibility in this case has had more than a century to prove itself and has managed to produce only a mountain of contradictions, to a degree where one can say with assurance that solutions do not lie with the local theory. The criticism is rather like objecting to Copernicus for widening our view of the universe from a cosy localism where problems might be settled simply to a cosmos so vast that problems could not be solved at all It may have seemed so in Copernicus' own time, but from five centuries on we can see that exactly the reverse has in fact proved to be true, at least for all the problems which could be conceived of in the fifteenth century.

The demand for all problems to be solved forthwith appears to be an outcome of the modern education process, in which students are subject to high-pressure packaging from the lecture-room to the examination-room, without them being given the opportunity to reflect decently on anything at all, a situation almost worse than the old days of illiteracy when at least there was an air of mystery and excitement about the world. R. L. Stevenson, despised among the literate with their preference for the pornograms of D. H. Lawrence, had it better when he said :

> O toiling of mortals ! O unwearied feet, travelling ye know not wither. Soon, soon, it seems to you, you must come forth on some conspicuous hilltop,and but a little way farther, against the setting sun, descry the spires of El Dorado. Little do ye know your own blessedness ; for to travel hopefully is a better thing than to arrive, and the true success is to labour

Much travelling remains to be done before the origin of life is understood, travelling we suspect in still unknown lands and continents.

P

Planets

P 1. Mars

An experiment performed in the Viking I landing gave a positive indication of the presence of life on Mars, the so-called Labeled Release (LR) experiment, whereas an organic analysis instrument (GCMS) gave a negative result. However, a subsequent test of both experiments with a soil sample from the Antarctic (in which life was known to be present) gave similar results, positive for LR and negative for GCMS (K. Biemann, *J. Mol. Evol.*,14, 1979, 65) so that the situation is now seen to be opposite to the way it was initially reported to the general public–the initial claim was that the negative result of GCMS disproved the presence of life on Mars. The claim was accompanied by reasons why the positive result of LR might have arisen from an unusual inorganic situation (as for instance the presence of hydrogen peroxide in Martian soil). This also has now been shown wrong, extensive attempts to reproduce the positive result of LR by non-biological means having failed (G. V. Levin and P. A. Straat, *Icarus*, 45, 1981, 494). In sum then, the Viking lander may well have detected life on Mars, but not with the definiteness one might perhaps have hoped. Nevertheless, the balance of the evidence is positive, a circumstance which has not been emphasised to the public. When big government science and the media blunder together neither is anxious to be seen correcting itself, a sufficient reason why nothing much that is good can come of government funded science done in the glare of publicity. Science is a quiet, reflective, essentially aristocratic activity, which

145

cannot flourish at all in egalitarian or totalitarian societies, and which is even being done to death by far too large government funding (with concomitant bureaucratic intervention) in Europe and the United States.

Humans have for long looked at the red colour of Mars and seen evidence there for the existence of life. The argument was that the red colour implied a highly oxidised condition, a conclusion that the Viking landings now seems to support. The supply of oxygen needed to produce this condition might have come either from photosynthetic organisms or perhaps more likely from iron-oxidising chemoautotrophic bacteria like those found by H. D. Pflug in the Murchison meteorite (entry M1).

The best chance for life to be currently active on Mars is probably inside glaciers, where it is possible for temperatures to rise sufficiently for water to become liquid. There would still be a problem of nutrient supply to microorganisms, but if the glaciers themselves turn over, top to bottom from time to time, this problem would be capable of solution. Bacteria in such a situation would need to live on some energy-producing chemical reaction, and, if the reaction had a gaseous product (CO_2 or CH_4 for example), the possibility would exist for the building up of subsurface pockets of gas, which might explode sporadically to the surface, unleashing quantities of spores, bacteria and inorganic dust into the Martian atmosphere. In this connection, a vast dust storm greeted the arrival of a Mariner vehicle in 1971, a storm that was attributed to high winds generated by the normal Martian meteorology, but if so one might wonder why such winds are not a regular seasonal phenomenon. The Martian event of 1971 was suggestive of the terrestrial jokullaups which occur every ten years or so on Grimsvotn, Iceland. These glacier bursts cause large trapped lakes which accumulated inside the glacier (due to volcanic heat) to break out with such violence that hundreds of square kilometres of land in the valley below become flooded, and vast blocks of ice are hurled far beyond the normal range of the glacier.

The Martian surface is cut by many sinuous channels which appear to have been made by a liquid much less viscous than molten lava. Water is a likely possibility, but since there is no body of liquid water at the surface of Mars nowadays, conditions in the past would need to have been considerably different from what they are today. Open surface areas of liquid water are likely to be of rare occurrence in the universe, a highly special condition suited to profound biological developments. If indeed the universe is biologically-controlled, such places would have a special and high importance. Although the thought is rather fanciful, the surface of Mars looks very much like a failed 'attempt', a failed 'experiment', of a kind which eventually succeeded in the case of the Earth.

146

P 2. Refractive Indices of Wet and Dry Bacteria

Under wet conditions about two-thirds of the volume inside a bacterium is water, which dries-out to leave a cavity under low pressure conditions in space and at very low humidities on the Earth. For bacterial spores under wet **conditions** about one-third of the interior volume is water (an inversion of the volume ratio of water-to-biomaterial from the bacteria themselves). Since a composite particle has a mean refractive index which in first approximation is equal to the volume average of the refractive indices of its components, the values for the various cases set out in Table P 2.1 are easily calculated in terms of n = 1.33 for water and n = 1.5 for biomaterial.

TABLE P 2.1

Refractive Indices of Bacteria and their Spores Under Wet and Dry Conditions, Compared with the Refractive Indices of Small Particles in Astronomical Sources

Case	Mean biological refractive index	Mean astronomical refractive index	Astronomical Source
Bacteria (dry)	1.167	˜ 1.16	Interstellar grains *
Bacteria (wet)	1.387	˜ 1.38	Jupiter **
Spores (dry)	1.333	Same as water-ice grains	
Spores (wet)	1.443	1.44 + 0.02	Venus ***

* For dielectric grains giving the visual extinction of starlight (entry 15).

** J. E. Hansen and A. Arking, *Science*, 171, 1971, 669. The upper clouds of Venus produce a rainbow, indicating that the particles of the clouds are, like spores, approximatley spherical with diameters of ˜ 1 um (see experimentally determined sizes in entry P3).

*** D. L. Coffeen and J. E. Hansen in *Planets, Stars and Nebulae Studied with Photopolarimetry*, ed. T. Gehrels, University of Arizona Press, 1974. The estimated particle diameters are ˜ 0.6 um, close to the peak of the histogram for the diameters of bacteria (entry B5).

Bacteria have remarkably broad tolerance ranges with respect to a wide variety of environmental factors (entry B1), tolerance ranges far too great to be explicable in terms of a supposed terrestrial evolution (entry B4). The properties of bacteria have a galaxy-wide quality, much wider than is needed for survival on the Earth alone. Indeed, because of the great breadth of their survival characteristics, the potentiality of bacteria to establish themselves on planets other than the Earth must be much greater than one would permit oneself to suppose within the narrow confines of an Earth-bound theory of the origin of life. This expectation is confirmed by the rather sensitive correspondence of refractive index set out in Table P 2.2. The presence of bacteria on other planets and on satellites is a natural expectation according to the present point of view, provided the body in question has an atmosphere in which microorganisms can land safely (entry A1). The latter conditions rules out Mercury among the planets, the Moon and other satellites without adequate atmospheres.

P 3. Venus

The distribution of particle sizes in the upper clouds of Venus is shown in Figure P 3.1, where by 'size' is meant the diameter of a circle with the same area as the silhouette of a particle (R. G. Knollenberg and D. M. Hunten, *Science*, 203, 1979, 792). The maximum of the figure corresponds very well with the maximum of the distribution of spore-forming bacteria given in Figure B 5.1, reproduced here as Figure P 3.2. On the other hand, Figure P 3.1 has a more extended upper tail. However, the distribution of Figure P 3.2 would also have a more extended upper tail if the rod-diameters of bacilli were changed to the mean silhouette diameters of Figure P 3.1. Thus a particle of length 3 μm and rod-diameter 1 μm would be included in Figure P 3.1 at an effective diameter of $2(3/\pi)^{\frac{1}{2}}$ μm but only at 1 μm in Figure P 3.2.

With this latter point in mind, the good correspondence of these figures suggests that the particles in the upper clouds of Venus are bacteria. A quite different view was arrived at some years ago by G. T. Sill (*Comm. Lunar and Planet. Labt.*,9, 1973, 191) and by A. T. Young (*Icarus*, 18, 1973, 564). Pure sulphuric acid has a refractive index of 1.48, but if ˜25 per cent water is added droplets of the mixture have a refractive index close to 1.44, the value required to explain the observed rainbow effect produced by the particles in the upper clouds of Venus (entry P 2). However, inorganic particles with a free surface (where condensation and evaporation can take place) are not usually of the same size distribution everywhere. Water droplets in terrestrial clouds vary by orders of magnitude in their sizes, as do the particles which condense in effluents from factory chimneys. Whereas particles of the same biological species maintain their sizes everywhere, inorganic particles vary markedly with the temperature, and also with the atmospheric partial pressures of the condensed substances. So here is a way to distinguish   between bacteria and droplets of sulphuric acids. Is the size distribution the same or does it vary markedly as altitude in the Venusian atmosphere is changed ? Figure P3.1 was for an altitude of about 60 km, while Figure P3.3 is for an altitude of about 40 km. The similarity evidently supports the bacterial hypothesis, which is also superior because it yields without hypothesis a refractive index of 1.44 for bacterial spores, whereas the 25 per cent admixture of water to sulphuric acid, needed to obtain a refractive index of 1.44, is an arbitrary amount.

The light reflected by the upper clouds of Venus is of a pale yellow colour, and this is also wrong for droplets of sulphuric acid which are colourless. On the other hand, bacteria, especially sulphur bacteria, could readily produce such a colouring. Thus at first contact with the facts the bacterial hypothesis wins three times over.

The particles at an altitude of about 55 km also contain a size distribution like those of Figures P3.1 and P3.3, but in addition there is a distinct and apparently separate distribution with its maximum centred at ˜ 2.5 μm. Since these other particles exist only

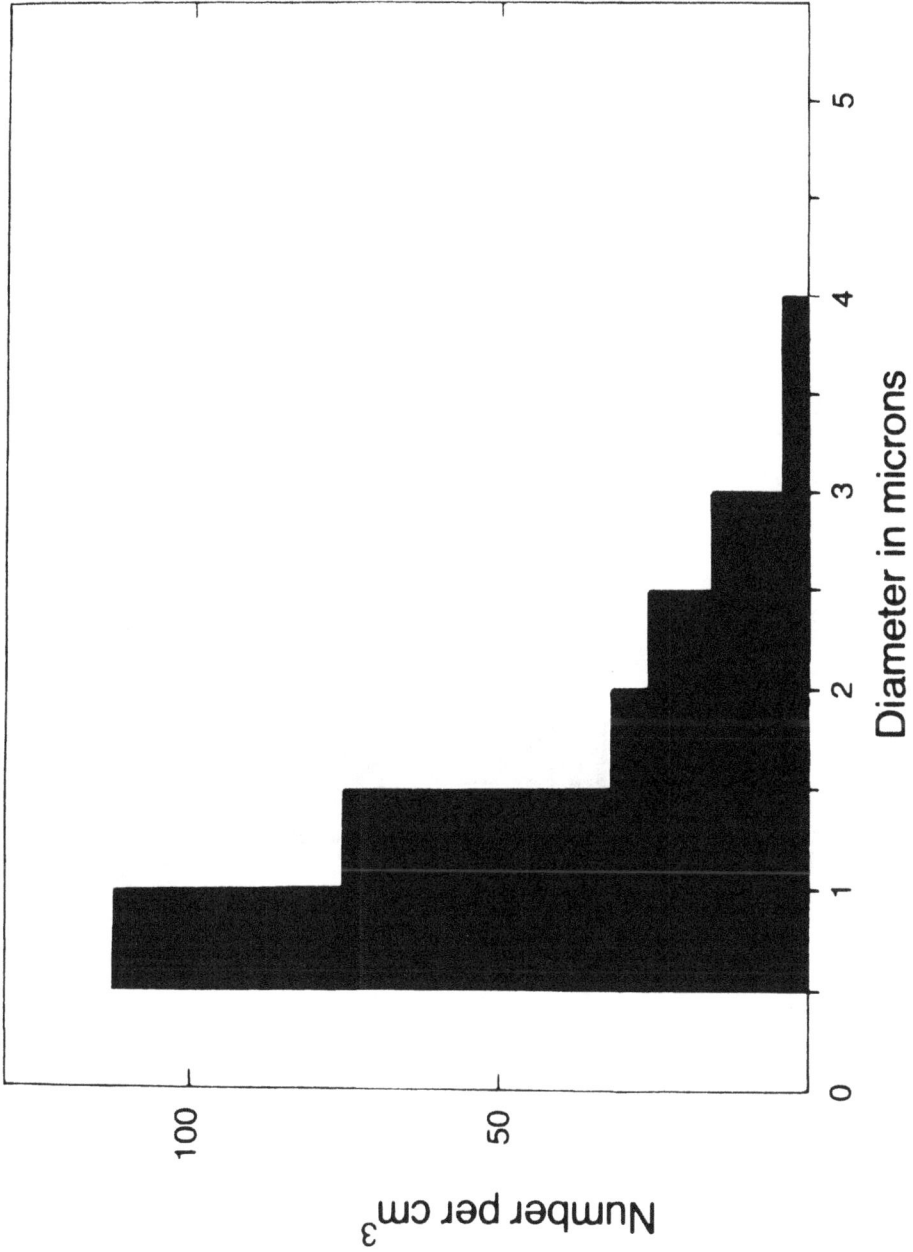

Fig. P 3.1.– Size distribution of aerosol particles in upper haze layers of Venus (R. G. Knollenberg and D. M. Hunten, *Science*, 203, 1979, 794).

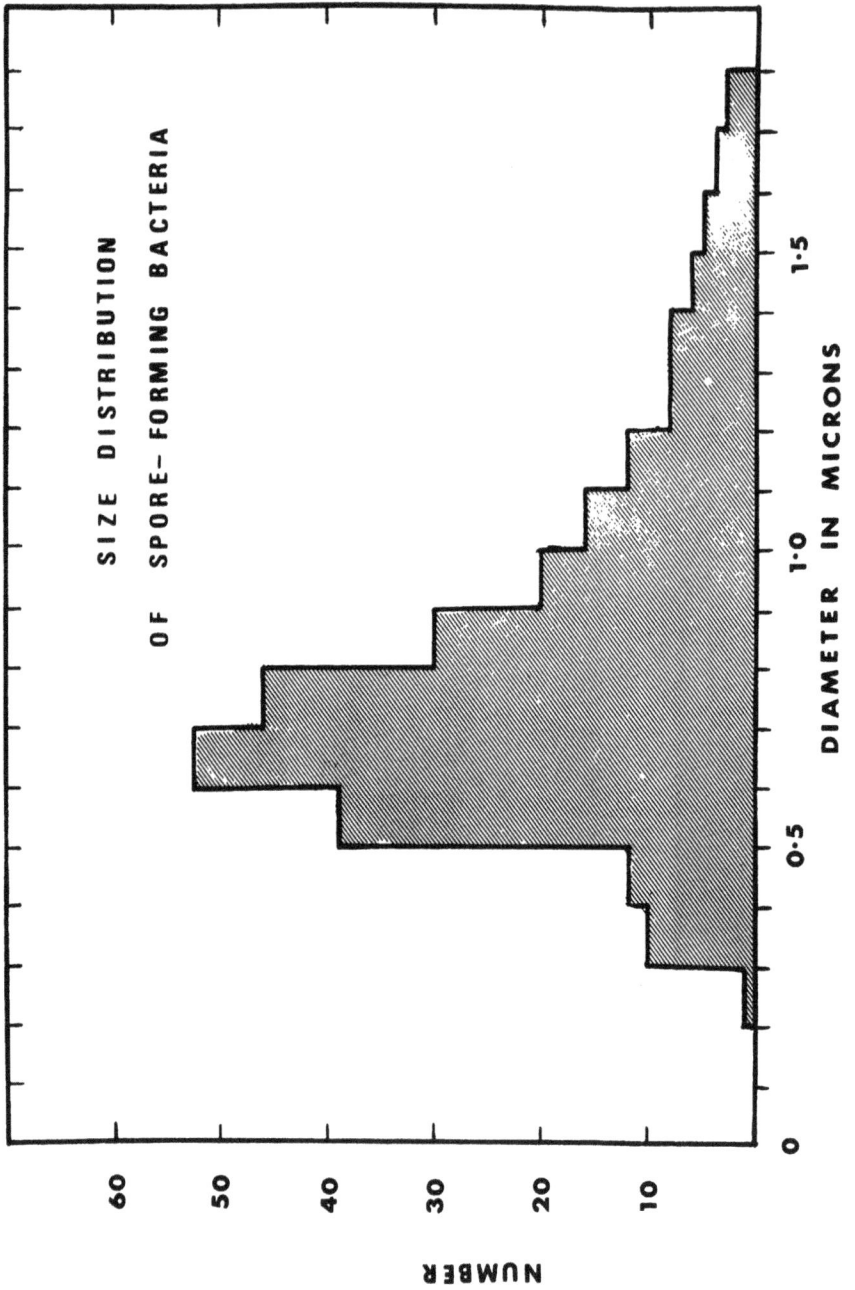

Fig. P 3.2.—Size distribution of spore-forming bacteria.

over a small range of altitude they could be inorganic, or they could be some protozoan or some group of bacteria. The usual view is that these larger particles (which comprise the main mass at altitudes ~ 55 km) are of free sulphur. Thus according to the usual view the atmosphere of Venus is a distinctly sulphurous place, clouds of acid droplets and of grains of sulphur itself. One would have thought it disturbing to this view that explicit measurements of the chemical composition of the atmospheric gases show remarkably little SO_2, as the following table of values obtained on 9 December 1978 with the Pioneer Venus sounder probe shows (V. I. Oyama, G. C. Carle, F. Woeller and J. B. Pollack, *Science,* 203, 1979, 802) :

TABLE P3.1

Atmospheric Composition of Venus as Measured by the Gas Chromatograph

Pressure (bars)	2.91 ±0.170	17.1±0.183
Concentration (%)	±confidence interval	(3 standard deviations)
CO_2	95.9 ±5.84	96.4 ±1.03
N_2	3.54 ±0.0261	3.41 ±0.0207
H_2O	0.519±0.0684	0.135±0.0149
Concentration (ppm)	± confidence interval	(3 standard deviations)
O_2	65.6±7.32	69.3±1.27
Ar	28.3±13.7	18.6±2.37
	+ 31.6	+5.54
Ne	10 − 9.6	4.31–3.91
	176 +2000	186 +349
SO_2	− 0	−156

A sample of the Venusian atmosphere was also taken at a total pressure of 0.698 ± 0.140 bars. but since the only real determinations at this higher altitude were for CO_2 and N_2 we have omitted this sample from the present table. In a subsequent communication (*Science,* 205, 1979, 52) Oyama *et al* discuss the calibration of their gas chromatograph and conclude that the best value for the SO_2 concentration is only ~ 180 ppm, from which they remark :

> These chemical properties of the Venus atmosphere make it extremely difficult to generate elemental sulfur particles either by gas-phase photochemistry near the cloud tops or by thermochemistry near their bottoms

Remembering that H_2O is not a common **constituent** of the Venusian atmosphere, the smallness of the SO_2 concentration is emphasised by the chromatogram shown in the above reference (obtained for the pressure of 1.77 bars at altitude 24 km). There may be sulphuric acid in the atmosphere of Venus, but if so it is likely to be the product of chemoautotrophic

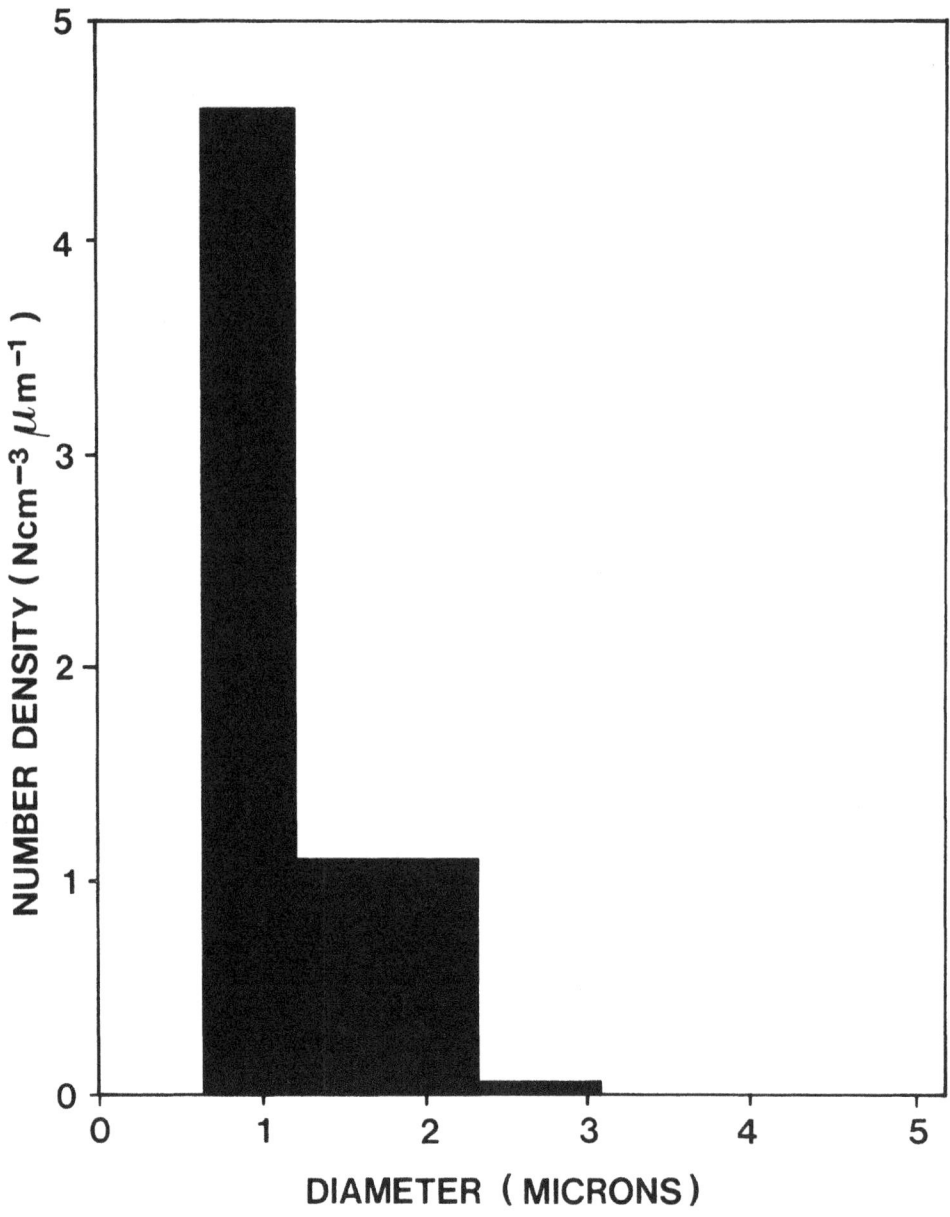

Fig. P 3.3. Size distribution of aerosol particles in ' lower thin haze region ' of Venus (R. G. Knollenberg and   D. M. Hunten, Science, 203, 1979, 794).

152

sulphur bacteria, not an inorganic product. When one looks at the facts, the usual sulphurous story has all the aspects of a chimera, which is to say a monster with a lion's head, a goat's body and a serpent's tail.

The bacterial hypothesis has its own problems to solve, however, the first one luckily not too difficult. Since Venus is exceedingly hot at ground-level (about 450°C) it would be impossible for bacteria in the atmosphere to survive if there were vertical motions all the way up from ground-level to the top of the clouds at a height of ~ 65 km. What seems to happen is that the Venusian atmosphere has a dual circulatory pattern. From the temperatures and pressures measured by Pioneer Venus, there is a region of stable atmosphere extending from an altitude ~ 35 km up to ~ 48 km which separates a lower very hot zone (where the temperature gradient is sufficiently close to the adiabatic condition for some circulation to be expected) from an upper much cooler zone which is actively convective from ~ 50 km to ~ 55 km, and with some degree of circulation very likely existing to the top of the upper clouds (A. Seiff, D. B. Kirk, R. E. Young, S. C. Sommer, R. C. Blanchard, J. T. Findlay, and G. M. Kelly, *Science*, 205, 1979, 46). Microorganisms would be unlikely to survive in the lower zone, because sooner or later they would be carried down to ground-level and cauterized there. On the other hand, microorganisms circulating in the upper zone would experience a highest temperature corresponding only to the altitude at the base of the upper zone, which for ~ 48 to ~ 50 km is ~ 80°C, well within the tolerance range for bacteria in Table B1.1.

The atmospheric pressure of 2.91 ± 0.170 bar at which the data of the first column of Table P3.1 was obtained **occurs** at an altitude of about 42 km, in the middle of the stable zone discussed above. This is the greatest height for which a definite determination of the water vapour concentration is available, viz., 0.519 ± 0.0684. Because of the ratio 44/18 between the molecular weight of CO_2 (the main atmospheric gas) and that of H_2O, a mass concentration of 0.519 percent implies a water vapour pressure of ~ 0.519 x 44/18 ~ 1.3 percent of the total atmospheric pressure, remarkably enough quite comparable to the situation in the terrestrial atmosphere. If this percentage were maintained upward from the 42 km level (where it was measured) into the clouds above 50 km, ice would condense on and within the cloud particles during upward convective movements to altitudes approaching 60 km. Thus the temperature falls to 0°C at about 58 km where the total atmospheric pressure is about 0.3 bar and the corresponding water vapour pressure would ~ 1.3×10^2 x 0.3 bar, equivalent to ~ 3 mm of Hg. Although the saturation vapour pressure of water at 0°C is somewhat higher than this, about 4.5 mm of Hg. a relative humidity of ~ 65% is sufficient to meet the requirements of some microorganisms (Table B1.1), i.e. for them to acquire water at this relative humidity. Since the temperature can be lowered below 0°C and bacteria still remain active (Table B1.1), possibly even down to −20°C where the saturation vapour pressure of water is less than 1 mm of Hg, the condition for microorganisms to acquire water can be satisfied *a fortiori* on Venus. Thus the temperature falls there to −20°C at altitude ∼ 60 km, where the total atmospheric pressure is about 0.2 bar, giving a water vapour partial pressure of 0.2 x 1.3 x 10^{-2} bar, equivalent to about 2 mm of Hg. Because only a relative humidity of 65 percent is needed in an extreme case (i.e. 65 percent of ∼ 1 mm of Hg at −20°C) a vapour pressure of 2 mm of Hg has a factor 3 in hand above what is strictly required.

In the present connection it is worth noting that there are many examples here on Earth of bacteria behaving as ice-nucleating centres. This is on account of their ability to hold water at vapour pressures below saturation values. Once ice has condensed on and within a bacterium, a high internal salt content can lower the melting point of the ice, so producing the liquid water necessary for microbial activity. This is the reason why bacteria can remain active below the normal freezing point of water (Table B1.1).

Although bacteria are small and able to ride easily with the motions above the intermediate stable zone (from ∼ 35 km to ∼ 48 km) there must nevertheless be occasional situations where a bacterium carried down to an altitude of ∼ 48 km fails to find an up-current on which it can ascend again. Inexorable gravity would then cause the bacterium to fall slowly down through the stable zone, until it reached altitudes below 35 km where vertical movements would soon snatch it downward to the high temperature of ∼450°C at ground-level. It is therefore to be expected that the stable zone will contain a thin haze of slowly falling doomed bacteria, and such a thin haze was indeed found by Pioneer Venus, with the particle distribution shown already in Figure P3.3.

From the data obtained by Pioneer Venus one can infer that the fall-out through the stable zone would denude the population of high-level organisms in upwards of 30,000 years, if there were no renewal of the bacteria. Water, nitrogen and carbon dioxide are available, but in addition to these main ingredients sodium, magnesium, phosphorus, sulphur, chlorine, potassium, calcium, manganese, iron, cobalt, copper, zinc and molybdenum are needed in smaller proportions. It can be calculated that the fall-out through the stable zone would carry about 10,000 tons of these essential ingredients each year from the higher convection zone to the lower zone. How may such a loss from the higher zone be compensated ? The meteoritic supply to the Earth is of about this amount (entry A2) and the supply to Venus should be approximately the same. This correspondence between the rate of loss of essential nutrients downwards from the upper circulatory zone on the one hand and a supply of the same elements from outside on the other suggests that the quantity of microorganisms in the Venusian clouds may be nutrient limited. It is interesting to contemplate what might happen if for some reason there was an interruption in the availability of one or more of the nutrients. The population of microorganisms would then decline until after a few tens of millennia it would become possible for an external observer to see down to the ground-level of Venus. There would then be a much greater penetration of sunlight to ground-level, and the increase in the

154

energy input, together with the formidable Venusian greenhouse effect, would drive-up the ground temperature much above its already high value of ~ 700K. A rise to the fusion temperature of the surface rocks could conceivably happen, causing any remaining volatile materials in the rocks to be evaporated into the atmosphere. From a biological point of view this dramatic process could serve the important role of maintaining an adequate supply of water in the atmosphere, not a great excess, but just adequate to maintain the shielding clouds of microorganisms, which is exactly how it seems to be—the water supply appears adequate by the sort of margin one would expect in a controlled situation, a margin that otherwise would seem **accidental**, which it almost surely is not.

www.ingramcontent.com/pod-product-compliance
Lightning Source LLC
Chambersburg PA
CBHW081516190326
41458CB00015B/5389